"FLYING BEATS WORK"

A Chuck Sassara's ALASKA

PROPELLERS, POLITICS & PEOPLE

Chuck Sassara (signature)

Cover photo – In 1954 Ann and I bought this brand new Volkswagen bus in Los Angeles, the first one we had laid eyes on. It looked perfect for our next adventure. We made it into a camper. The VW dealer in Seattle was delighted to service it for us since he had not yet seen one. After 2,200 miles of rocks, dust and mud on the ALCAN Highway, we rolled into our new home, the Territory of Alaska, on June 20, 1955.

Library of Congress Control Number: 2015939186

ISBN: Hard Cover: 978-1-57833-616-6
        Soft Cover: 978-1-57833-617-3

Book design: Vered R. Mares, **Todd Communications**
All photographs: from the author's personal collection and used by permission
Editors: Flip Todd and Carmen Maldonado

Printed by Everbest Printing Co., Ltd., in Guangzhou, China, through **Alaska Print Brokers**, Anchorage, Alaska

Published by:

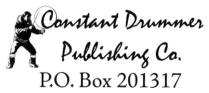

**Constant Drummer
Publishing Co.**
P.O. Box 201317
Anchorage, Alaska 99520
charlesjonessassara@gmail.com

Distributed by:
Todd Communications
611 E. 12th Ave.
Anchorage, Alaska 99501-4603
(907) 274-TODD (8633) • Fax: (907) 929-5550
with other offices in Juneau and Fairbanks, Alaska
sales@toddcom.com • **WWW.ALASKABOOKSANDCALENDARS.COM**

Dedication

To Ann…

…my wife of fifty-eight years, a woman of grace and warmth and wit who never wavered in her loyalty, no matter which way our trail turned. She was the love of my life.

She died in my arms as I held her upright, trying to squeeze one more moment of life into her. Flashbacks raced through my head.

Ann at the wheel of a 40 foot ketch in pitch black seas. She was looking up selecting the stars she would navigate by as we passed through the straits.

Ann sitting on a log by a fire in the jungle listening to a young black woman who was explaining how she was facing the future. The woman's two young daughters were huddled tightly to Ann, clutching her sleeves. I had stepped back out of the circle of light so they couldn't see the tears welling up in my eyes.

Ann holding the door open for any weary traveler.

Table of Contents

"Chuck, I'm going to teach you how to stay alive. Always fly looking back for a way out of trouble ahead. Don't take off if you merely *think* you can complete the flight. Take off only when you *know* you can. I hold the unchallenged world record for 180-degree turns and I want you to be in second place."

— *Bill Borland,*
Chief pilot, Reeve Aleutian Airways

Chapter 1
Chakachamna Chicken Soup

Bethel, Alaska, 1976

Keeping aircraft operating in the Alaska Bush is a constant headache and juggling act. At Christiansen Air Service, where I worked a couple of years as chief pilot, we maintained and operated three Cessna 172s, two Cessna 180s, two Cessna 185s, a 206, a 207, a Pilatus Porter, and an Aero Commander 560. Parts and pieces were switched from one airplane to another, as needed, but sometimes we had to fly a plane into Anchorage for repairs.

On January 16, 1976, I told the gang I was taking the 206 into Anchorage for installation of a radio. I would be back in a day or two. A check of the weather revealed little to worry about on the most direct route, a four-hundred-mile flight with two mountain ranges and lots of rocks and tundra to look at. Crossing the Alaska Range, I would fly through the 3,176-foot Merrill Pass, then let down and run the length of Chakachamna Lake and on to Anchorage.

I'd made this run many times and, as always, filed a flight plan with the Federal Aviation Administration's local Flight Service Center. Hearing of my plans, Dave Fitzwater, a local drifter known as "Fish Water," showed up at my office wanting to bum a ride into Anchorage.

"Sure," I said, "throw your gear into the 206 and let's get out of here."

As we flew over the remote Sparrevohn Air Force Station, a radar site about half-way to Anchorage, I looked up into Merrill Pass ahead where the narrow flight path does an S-turn through the mountains. The weather looked good, but you never knew for sure about conditions at the other end of the pass once you had committed.

After making it through the pass, I was flying tight against the mountainside on the south side of the lake, following the shoreline. Snow began to fall, but this was not a concern at first. By the time I reached the far end of the lake, however, snow was coming down hard. The door ahead was slamming shut, so I made a 180-degree turn and headed back to the west end of the lake. By then, heavy snow was beginning to fall there, too.

I thought about flying back through Merrill Pass to Bethel and putting off my business in Anchorage until another day. From my perspective, the pass ahead looked inviting, but I had no idea about weather beyond the S-curve. If it had deteriorated behind me, I risked getting stuck in the pass with nowhere to go. I thought about the people who had died here and about the wreckage of fifty or so aircraft that decorated the pass. Just the year before, a twin-engine Evangel 4500 flying blindly in bad weather had crashed into the rocks, killing the pilot and a load of sled dogs.

I knew these wrecks had two things in common: the pilots believed that, for whatever reason, they *had* to get through, and they died trying.

So, we executed yet another turn-around and headed east again. The slopes on either side were steep, rising to 11,000 feet on the north side and 7,700 feet on the south. This was no place to run out of flying room.

Soon we were hemmed in on all sides by a serious snowstorm. Visibility was disappearing rapidly, and the Cessna's right wing tip was only about a hundred feet from the rock face. I got the message: the weather gods were telling me to put the 206 down and walk away.

Dave Fitzwater looked terrified. He realized this routine flight was going to hell in a hurry.

We needed to get down. I chopped the power, dropped twenty degrees of flaps, and made a slight left turn out toward the middle of the frozen lake. We descended into a whiteout with nothing but snow to be seen up, down, and sideways. Slowly … slowly … CONTACT. The left main gear touched first, instantly burying itself in deep snow, spinning the aircraft around ninety degrees, and the left wing tip and nose gear slammed down hard.

We were alive and unhurt.

We stepped out of the plane and looked around. There was not much to see. It was snowing furiously. The Cessna's prop was bent and its left wing tip mangled. The landing-gear leg and nose strut were okay but two wheels were nowhere to be seen. All I saw were tracks leading away into the pale distance. I followed them.

"Hey, Dave," I yelled, "you won't believe this."

The Cessna 206 wouldn't be going anywhere soon. The four bolts holding the axle onto the gear leg had snapped off, as had four bolts that attached the fork to the nose strut. When the plane stopped abruptly, two tires and axles kept going another hundred feet or so. We picked up the wheels and tires and carried them back to the plane. It seemed like a good idea at the time. Later, it turned out to have been a *great* idea.

I crawled into the rear of the cabin to turn on the Emergency Locator Transmitter. *Son of a bitch, no ELT.* The mechanic had pulled it out for use in another aircraft. As chief pilot, my responsibility was safety, and I cursed myself for not having checked to verify that we carried an ELT. All we had with us were a suitcase full of dirty laundry, one-quarter of a tin of peanut butter, and the knowledge and will to survive.

Just for the hell of it, I tried calling on the VHF radio that hadn't been working. No luck. Replacing the malfunctioning radio had been the reason for flying into Anchorage.

I asked Dave to pull out the seats so we would have room to make a bed in case we had to spend the night. What an optimist I was. While Dave set about this task, I brushed snow off the fuselage, which was painted maroon and would stick out like a sore thumb in the white world surrounding us.

The temperature hovered at zero degrees Fahrenheit—just right for a campout. Our first priority was to build a fire for warmth and to signal planes overhead. Being the scrounger that he was, Dave brought back a two-foot-long stick, the only wood to be found nearby in the rocky lake basin, which was covered in a thick blanket of snow. Pickings were slim.

Knife in hand, I slowly dismantled a few things that might prove useful to our survival. First, I hacked off from an outer wing panel a piece of aluminum skin, which I carefully placed down onto the snow and stomped into the shape of a bowl. Then I removed the cargo straps and seat belts. The cargo door became a sled.

The plane was leaning over sideways. We needed a level place for sleeping. So, I stabbed the air out of the remaining tire and tube on the passenger side. Problem solved. Then I slid the knife into a seat-back and pulled out the sponge-rubber cushion, the makings of a bed.

By nightfall, the expectation that we would be picked up right away had faded. Instead, we spent the night tending a fire consisting of one part tire, one part inner tube, a splash of engine oil, and Dave's stick. The solid overcast kept away air traffic. Our faces turned black from sooty smoke pouring out of our fire, but we kept warm by stomping our feet and taking turns swearing.

A new day came with the same lovely weather. The elevation of the lake was about 1,100 feet, so our heads literally were stuck in the clouds. I stretched, scratched, brushed my teeth, and began a more serious assessment of our situation. We had a good shelter from the wind, snow, and rain, so there was no reason to wander away from the plane. We would be spotted easily when the weather cleared.

A few years before, an inexperienced pilot had crashed on the snowfield above us. Apparently he had seen the lights of Anchorage about sixty miles to the east. He left a note ("I'm walking out.") on a seat in the plane. A day or so later, searchers found the note, then followed the pilot's tracks for about a mile to a precipice. It looked as if he had walked off the edge. The powdery snow far below covered his final resting place.

Though I had no intention to leave the plane, I made a pair of crude snowshoes using a piece of seat, seatbelts, and tie-downs. Wearing the snowshoes, crude as they were, was easier than slogging around in deep snow.

Dave was tall and thin, and the cold was starting to get to him. He stayed in the plane huddled under the engine cover, normally used to wrap around the engine to preserve heat when starting it in cold weather.

By nature, Dave was gregarious, but he stopped talking on the second day—a bad sign. I reminded him that we were on a busy flight path and would be found when the weather cleared. I urged him to get out and move around. Having at least forty gallons of fuel and a plumber's blowtorch on board, we could start a hell of a fire.

My pep talk had little effect, so I kicked Dave out of the sack and put him to work cutting up the two remaining tires and tubes. I figured this would give us fuel for ten small smudge fires.

The Cessna's landing light resisted our salvage effort for a while, but finally we extracted it from the wing along with some wire. Now we had a good signal device. But I needed a power source, so I pulled out the battery.

Around eight-fifteen p.m., after connecting the battery to the landing light, I lay on my back in the snow looking straight up hoping to spot the lights of a passing airplane. At the first sign, I was ready to project our distress beacon into the sky. A Wien Consolidated Airlines jet on its way to Bethel passed over the lake about this time every evening. I figured news of our disappearance had been circulated widely by then, and anyone seeing a light on Chakachamna Lake would guess it was me. But there were no holes in the cloud cover. Maybe tomorrow.

On the fourth day, Dave began to worry me seriously. He didn't look well and seemed morose. He was given to sighing, rubbing his head, and mumbling to himself.

"Hey, pal, what's the matter?" I asked.

"I don't know what to do," he said.

"About what?" I asked.

"I don't know whether to make it long and skinny, or short and fat."

I was mystified. "What the hell you talking about?"

"You won't understand because you don't smoke," Dave said. "I only have enough stuff to make one cigarette." I knew then he was not going to fall apart on me.

Another piece of aluminum we fashioned into a stove of sorts with the plumber's torch blazing away on a Sunny Jim peanut butter can. Now we had a supply of hot water, which would ward off hypothermia. The plane was actually rather comfy.

Meanwhile, I kept up my vigil, listening and scanning the sky. Most of the time our surroundings were eerily quiet.

The fifth day was dull and uneventful. Not even a bird came by. I was tempted to put on my home-made snowshoes and take a look around. I recalled there were trees somewhere to the north. But instinct and training told me to stay with the plane.

Dave did not respond to my cheerful one-sided conversation. He just wanted to sleep, but the cold was getting to him, and he wanted to fire up the blowtorch inside the plane. To preserve the fuel, I made him keep moving and drink hot water instead. He would be all right. Of course, this was easy for me to say—I was making out fine with my down jacket and extra layer of natural insulation.

"We are going to get out of here soon, pal," I promised.

By nine a.m. on the sixth day, the ceiling was starting to lift—not much, maybe fifty feet—but at least we could see down to the east end of the lake. This was the direction from which help would come. I swept off the plane again. Then I opened a can of oil and drained some gasoline from the wing tank. We were ready to fire up the smudge pot. The ceiling continued to rise, as did our spirits.

"Hot damn, Dave, I hear an airplane," I shouted. "It sounds like a de Havilland Beaver with a Pratt & Whitney 985 engine coming up the draw."

As the drone of the approaching aircraft grew louder, Dave suddenly came to life, dancing a jig while whooping and whistling. I laughed wildly at his antics while pouring gasoline onto the pile of rubber and torching it. The black smoke began to rise, angling slightly—perfect. A black line coming off the surface of the lake would look like a giant checkmark on a piece of white paper.

The beautiful Beaver came rumbling straight at us, making a pass so close I could see Gene Wieler at the controls and my beautiful wife Ann waving from the back seat. They made one circle and then headed east again where the creek flowed off the lake, winding its way down into the flat country.

I knew the protocol in this situation would be to notify the Rescue Coordination Center at Elmendorf Air Force Base, providing a position of the downed aircraft. The RCC would arrange a pickup. Sure enough, about an hour later, we heard the sweet sound of a Continental I/0 520 engine, and soon a fire-engine red Cessna 185 on skis appeared.

The pilot landed next to us and extended an invitation we could not refuse.

"Get in," he said, "I'll take you home."

Once we were belted in, he poured on the power and away we went. I looked around at the rapidly clearing skies. Damn, those mountains were beautiful.

I assumed we were headed for Anchorage, so I was surprised when the pilot pointed down to a little strip on the edge of Cook Inlet. There, a Civil Air Patrol Beaver sat waiting. For the first time I was overwhelmed with emotion. There was my beloved wife and my brother Dick. It was a reunion I will never forget. We piled into the Beaver and blasted off across the inlet to Merrill Field, rolling to a stop in front of the CAP hangar, where a small crowd greeted us—another surprise.

Amid the laugher, handshakes, and hugging, a young journalist stuck a microphone in my face and asked how I felt.

"Fantastic!" I answered.

"What did you do to survive," she asked.

"Ah, we just sort of hung around mostly," I said, a bit taken aback.

"What did you eat?" she asked.

Without a thought, I said, "Chakachamna chicken soup." This seemed to be as good an answer as any.

"What's that?" she asked.

"That's where you put some snow into a Sunny Jim peanut butter can, imagine yourself adding a piece of chicken, bring it to a brisk boil, and then drink it quickly before the illusion fades."

My fanciful quote was reported widely by the Associated Press. I received calls from news media all over the country, including the *National Enquirer*, which offered $1,000 for an exclusive story. I passed.

When we got home, Ann urged me to call Ray Petersen, the popular former Bush pilot who ran Wien Consolidated Airlines. "He perked me up when I was deathly afraid for you," Ann said.

"Welcome back," Ray said when I called. "How did it go?"

"Pretty good," I said. "You know Merrill Pass, it can bite you in the ass."

I thanked him for consoling Ann and giving her hope. Ray had predicted we were sitting out the weather somewhere. "It sure made her feel a lot better coming from you with all the hours you spent flying back and forth through that country," I said.

"I told her you were probably on Chakachamna Lake, as I was one time during World War II," Ray said. "I put a plane down there with a USO troupe on board—a bunch of girls. We had a hell of a party. Anyway, glad you made it."

I told Ray I had shined a light straight up into the night sky trying to attract attention. "I knew you had a Boeing 737 scheduled to fly over about eight-thirty p.m., so I tried to light up the cockpit by shining it at them and rocking it back and forth."

"That proves one thing," Ray joked. "You can't see a landing light through a *Playboy* magazine."

Of course, I knew Wien pilots on the Anchorage-Bethel route, most of whom I knew, had been looking for us the past five and a half days.

The next call was from a friend, Al Crook, head of the FAA in Alaska. "You've got to come to my office and file an accident report," he said.

"Gee, Al, this wasn't an accident, only an incident," I said, but he insisted.

When I walked into Al's office, he handed me a blank form. "Here, fill this out so I can get back to work. The phone has been ringing off the hook—reporters from back East, the White House, senators, congressmen, and big-wigs with the FAA in Washington—you name it. They want to know what's going on. You made too many headlines."

I hadn't known any of this. So I sat down at Al's desk and began to fill in the blanks. Name: yep; date of accident: yep; type of aircraft: yep;

location of accident: yep. I wrote "N/A" in the other blanks.

At the bottom was room for a short narrative. "Continued flight under these conditions would have resulted in my death," I wrote, then signed the document. Copies were sent to all who had inquired. We never heard another word about it.

Working with my insurance company, I arranged for a helicopter to airlift the 206 to Kenai for repairs. There was little damage beyond the bent wingtip. Of course, we needed six new seats, a propeller overhaul, a new set of tires, and a patch of replacement skin for the wing.

The insurance company balked at the cost of replacing the seats, noting that I had destroyed them myself. My insurance agent, Lois Clary, stuck up for me. She argued that my use of the seats as cushions for sleeping may have saved the life of my passenger, pointing out that in Alaska the insured party has a legal obligation to give immediate medical aid to anyone injured or at risk of injury or death.

To her credit, Lois hammered away at this until the insurance company backed down.

* * *

A MONTH OR TWO later, I called Carol Hansen, who had been my teammate on the football team at University High School in West LA.

"Hi, Carol, this is Chuck Sassara."

"WHAT! Who is this?" he said, his voice rising in disbelief.

"Chuck Sassara," I repeated.

"You're supposed to be dead."

"I'm not dead. Where did you get that idea?"

"I was driving along the Pacific Coast Highway listening to *Paul Harvey News* on the radio," Carol said. "He said you were lost in an airplane in Alaska and presumed dead. This shook me up, and I told all your friends what happened."

"Well," I said, "I sat out there in a snowbank for a week, then got picked up. I got hauled back to Anchorage, kissed Ann, got a haircut, and went back to work flying."

I wished Paul Harvey had delivered on his daily promise to report "the rest of the story." But he didn't, so I'll tell you the rest of my story myself, from the beginning.

Chapter 2
A Family on the Move

Pittsburgh, Pennsylvania, 1910

My grandfather, Charles Jones Sassara, was a stage actor turned circus clown who fell in love with my grandmother, Sophie Belle Rhay, who had been Miss Pittsburgh. He persuaded Sophie to run off with him, and that meant joining the circus, too. He was thirty-five. She was sixteen.

Circuses traveled by train in those days. This one was heading east when it made an unscheduled stop, unloaded the whole crew, and set up picnic tables. Everyone cheered enthusiastically as my grandparents exchanged marriage vows on a summer day in the Pennsylvania countryside.

Little Charlie, my dad, was born a year later. My grandfather died when Dad was only two. Dad's mother Sophie remarried, but died on an operating table during what was supposed to have been minor surgery. Dad was seven.

Charlie turned out to be a handful for his stepfather, Frank Shimpton. My father was an independent little cuss who kept running away from boarding school, prompting Shimpton to deliver young Charlie to Pittsburgh to be raised by Sophie's mother. Charlie loved his grandma and two uncles. His favorite uncle was John Sims, a full-time truck mechanic and part-time professional boxer who had fought in more than a hundred bouts. Uncle John taught my father how to fix things and how to defend himself.

Charlie's grandmother was having a hard time making ends meet, in part because her husband had a sorry habit of drinking up his paychecks. Charlie, by then thirteen, could see he was a burden, so one day he jumped on a freight train headed for New York City. From there he stowed away aboard a freighter, but was discovered and made to peel potatoes all the way across the Atlantic. In Bremen, Germany, he found work building crates for ocean-going freight. Four months later, he heard about a ship soon to depart for South America. Once again he scrambled up a mooring line and stowed away. Next port of call: Barranquilla, Colombia.

And this is where my Dad's extraordinary South American adventures began.

* * *

AS CHARLIE WALKED along the Calle Dos waterfront on his way to Felix's bar, where a five-centavo beer awaited to slake his thirst, he returned the waves and friendly calls from the women who were beginning their nightly routine of selling sexual favors to sailors and dockhands who frequented ships' stores, warehouses, raucous bars, and flophouses. Horse-drawn wagons loaded with freight plodded up and down the muddy, pot-holed streets. Stray dogs lounged in the shade, keeping an eye open in case someone dropped something edible. Seagulls patrolled overhead, always ready to open their bomb-bay doors and drop a load of white excrement on the people below.

Life was good. Charlie had gotten a job building wooden shipping crates for seventy-five cents a day. His boss allowed him to sleep in an empty box at night.

In front of Felix's bar, a young girl was frying handmade empanadas stuffed with ground chicken or iguana and potato filling, which she sold to the passing crowd. Charlie smiled at the pretty girl, handed her five centavos, and took his meal into the bar.

Barranquilla was a raw, sometimes lawless town, and Charlie always made sure to sit with his back to the wall. He nodded to the bartender, who slid a cold beer his way. The crowd of regulars was just beginning to show up. The noise level rose and the smell of sweat-stained clothes filled the air. The only air-conditioning in a South American bar in 1925 was the constant swinging of the front door.

As the evening wore on, after considerable drinking, a fight broke out and the place was suddenly filled with men brandishing knives and broken bottles. Charlie thought he heard someone say in Spanish, "Let's rob the American kid." He took off running.

Wearing cheap patent-leather shoes and a thin shirt and pants, my father walked west out of Barranquilla following a trail into the nearby jungle. He decided he was finished with the port city and kept right on going, finding a few mangos to eat along the way and sleeping under trees at night. He kept going, even after his shoes fell apart. The ground changed from mud to a well-packed dirt trail as he climbed into the Andes Mountains. Nothing edible grew alongside the trail. At one point he subsisted five days on a dried-up orange peel.

High in the mountains, crossing over into Ecuador, my father followed a constantly twisting trail for nearly a thousand miles to Guayaquil, Ecuador, always on the lookout for ways to feed himself and earn a few pennies along the way. He celebrated his fourteenth birthday by eating a bunch of bananas. He found work for a while digging ditches for a company building a water pipeline.

A couple months later, Charlie was off again, stowing away on a ship headed for Santiago, Chile. From there, he worked his way back over the Andes into Argentina.

After a decade of wandering back and forth across South America while working all kinds of construction jobs, Charlie decided it was time to go home and see his grandmother. He was twenty-three.

On his first day back in Pittsburgh, my father ran into a fellow named Angelo, whom he had known in grade school.

"Hey, Charlie, where have you been?" the former classmate asked.

Charlie recounted his adventures. Angelo was suitably impressed. Then the topic turned to a more important subject: girls.

"I know two cousins, Anna and Catalina," Angelo said. "They are a lot of fun. You want to go on a double date?"

My father dated Catalina, a pretty, dimpled woman with red hair. Just as his parents had fallen in love at first sight and married quickly, Charles Sassara Jr. and Catalina LaRocca soon went looking for a justice of the peace office in downtown Pittsburgh. They paid two dollars for a marriage license.

The marriage lasted fifty-nine years. My mother, who took the name Kathleen, an Americanized version of Catalina, had an adventuresome spirit—just like dad, maybe even more so. Those two were ready to do anything, go anywhere, anytime. Among family and friends they became known as the "gypsies."

Uncle John helped my dad get a job as a bus mechanic in Detroit, Michigan. I showed up on October 19, 1930, born Charles Jones Sassara III. Six months later, Dad arranged to deliver a new Ford Model A from the Detroit factory to a buyer in Los Angeles, California. Off my parents went with the baby (me), a trip that was the start of my life as a wanderer.

America was stuck in the Depression. Dad had been laid off. In LA, Mom waited tables while my Dad stayed home taking care of me and Dick, who came along in 1933. Our father baked pies, sometimes as many as twenty-five a day, which mother sold to the restaurant where she worked.

But Dad was restless.

Chapter 3
Life in Florida

Miami, 1935

"Kathleen," said Dad (whom we also called "Pappy"), "let's pack up and go to Florida. I think I can find work there."

I remember going into a hardware store in Los Angeles with Dad to buy a tent before we piled into a Willys 77, the cheap little car of the day. In Miami, we lived in a campground until Mom found a residential lot for ninety-nine dollars down and nine dollars a month. Dad went to work as a lineman for Florida Power & Light. When the wages began to flow, my parents built a home.

Things were looking up—that is, until a man showed up with a letter from the bank in California that had financed the Willys. Apparently in those days you weren't supposed to take a car out of the state in which it was financed, if the car had a lien on it. My mother had been mailing in payments on time, but when the California bank saw the Florida post-marks, it repossessed the Willys. Our next car was a big old Hudson sedan equipped with window blinds and flower vases built into the back seat.

My first airplane ride was in the back seat of a tandem something or other in 1936. I was six and my brother was three. Dick and I held on to each other tightly as we floated around over downtown Miami. The ride cost three dollars, a princely sum then.

My main concern about Miami was that it lacked snow for Santa and his reindeer at Christmas. My mother assured me Santa had made special arrangements to fly to Miami in an airplane. Problem solved. We lived near the Miami airport. For weeks before the big day, I minded my manners, ate all my dinner, and brushed my teeth. But mainly I hid under my blanket pretending to be asleep whenever I heard a plane flying over at night. Santa Claus is quite vigilant, you know.

My second flight took place in 1940 in a four-engine Sikorsky fly-ing boat that made regular stops at Dinner Cay, near Coral Gables. I was inspired by this thrilling flight to Havana, Cuba. At the Pan American World Airways terminal, Mom bought three tickets – hers for 20 dollars, Dick's and mine for ten dollars each. Part of the boarding process was to climb onto a large scale so Pan Am could record your weight. In those days, the captain demanded a full accounting of every pound he was to carry.

Near the end of the dock, we climbed a portable stairway to the top of the fuselage, which sat well down into the water. I remember descending into the passenger cabin. Taking my seat, I glanced out the window. Holy smokes! The water outside was nearly eye level. Soon we were pushed away from the dock. The line crew wore bathing suits.

Although this was my first ride in a big airplane, I was familiar with the Sikorsky, having seen it land and take off many times. We often went out to Dinner Cay on Sundays, sat on the grass with our picnic lunch, and watched the activity.

Nothing compared to seeing those four big, round piston engines being fired up, one by one. Each was cranked over and came to life belching black smoke, blue smoke, and fire. It was an awesome display of the engines' brute power. How could a nine-year-old kid not imagine himself becoming a pilot?

My heart pounded as we taxied out into Biscayne Bay. The captain ran up the engines, turned into the wind, and poured on the power. A roar echoed across the bay as 4,000 horsepower went to work, and the Sikorsky leaped forward. Water flew everywhere as we began rising to the surface, picking up speed quickly. Suddenly, we lifted free—we were flying. This was a rich, indelible experience that left me feeling we could go anywhere in the world. There was a reason Pan American operated those huge flying boats: places like Rio de Janeiro, Honolulu, and Singapore had no airports then.

The flight to Havana hooked me. I was determined to fly.

From Havana, Mom, Dick, and I transferred to a United Fruit Company ship that supplied the Panama Canal Zone, where Dad had found a new job. Our three-day voyage was yet another adventure for the Sassara boys.

In Panama Dad met us at the dock. "You guys will love this place with its banana plantations, iguanas climbing around in the mango trees, and great fishing."

Boy, did we ever love it, especially my mother. Without hesitation, she jumped into the Zone's social world and soon became well-known as an activist, always ready to organize parties or parades. She never said no to the various preachers who showed up on our doorstep, looking for help raising funds.

"Heck, yes, just leave it to me, I will get it done," Mom said.

"That's great, Mrs. Sassara, can we expect to see you in church on Sunday?"

"No," she replied, "that is the day I go fishing."

Life was sweet in the Canal Zone. As a newly minted Boy Scout, I went on camping trips to jungles that would have seemed alien to scouts in the States. The boys in our troop carried machetes and twelve-gauge shotguns, clutching the latter to our chests as we slept in hammocks at night. All kinds of animals ran under and around us in the dark.

In Panama, prostitution was legal and rampant. A twelve-year-old could buy a shot of whiskey, if he was tall enough to slap his money down on the bar. But I had better things to do. I was busy learning how to build boats. The first was a twelve-foot rowboat to which I mounted a keel and added a sail rig. Dick and I charged across Gatun Lake, taking care to stay out of the canal's shipping lanes.

* * *

OUR PERFECT LIFE was interrupted on December 7, 1941.

Dad and I were working on the brakes of his 1937 Dodge sedan on a typically sunny morning when the phone rang. I answered it.

"Is Charles Sassara there?"

"Yeah, sure, I'll get him, he's my dad," I said.

"Well, hurry along—it's very important that I talk to him."

"Hey, dad, some guy wants you on the phone right away, says it is important."

My dad crawled out from underneath the car, stopped long enough to grab a clean rag and wipe his greasy hands, then picked up the phone.

"Sassara here," he said. I couldn't hear what was said but my dad's answers were clipped and to the point.

"Yes," he said, a funny look on his face. "I understand. I will take all necessary steps."

"We are at war, the Japs have attacked Pearl Harbor in Hawaii," he told me after hanging up. "The man on the phone told me to do whatever needs to be done to protect the electrical grid that serves the canal."

I ran over to the large warehouse that was temporarily housing a hundred or so National Guard soldiers who had arrived in the Canal Zone the week before. The sentry out front was sitting in a chair, tilted back against a wall, his rifle leaning against it nearby.

Unable to contain my excitement, I told him, "You better get ready, there's a war on! The Japanese have attacked the U.S. Navy base in Hawaii."

The soldier opened one eye, saw an eleven-year-old kid jumping up and down, then closed his eye and went back to catching the morning sun.

In Panama, bedlam occurred as the U.S. military went on a war footing and prepared to defend the canal against Japanese invaders. We kids quickly mobilized our defenses, too, carrying wooden rifles and bottles of water to the highest ground around the Gatun Locks. There, we dug foxholes with a clear line of fire over the locks. We were the first defenders to arrive. An hour later, a sweaty Army sergeant struggled up to the top of the hill, followed by a squad of soldiers lugging forty-millimeter antiaircraft guns. The guns were placed in our carefully selected positions.

My dad, an electrical engineer, applied for a leave of absence to join the military, but he was told his work in the Canal Zone was too essential. He asked again and again until finally he was allowed to enlist in the Navy in late 1942. He was sworn in as an electrician's mate first class, having turned down an opportunity to go in as an officer.

Suddenly Mom, Dick, and I became military dependents and were ordered to leave the Canal Zone immediately by the first available transport, which turned out to be a troopship carrying wounded to New Orleans. We were headed for Pittsburgh, where we would wait out the war. Our ship joined a small convoy escorted by two destroyer escorts. The voyage was expected to take three or four days, but was prolonged after a German submarine began following us. Dick and I sat on the fantail watching a Navy gun crew fire its three-inch cannon at the sub's periscope as our ship rolled from side to side. The first round overshot the target. The second fell short. The gun crew must have fired ten shells, but none hit the periscope or came close.

Soon, one of our destroyer escorts came flying in from the horizon, doing about forty knots, and charged right over the top of where the submarine had last been seen. Depth charges were pushed off the stern, the explosions lifting the destroyer almost out of the water. The war ship did a quick U-turn and made a second pass. By then, the periscope had disappeared, and the sub had taken evasive action. But another six or seven depth charges were lobbed out for luck.

Next, two anti-submarine amphibian aircraft showed up. They flew around for a while, searching. The water in this part of the Caribbean was quite clear, so I believe the sub may have been spotted. A rumor spread throughout the ship that it had been sunk by the destroyer.

Our convoy plodded along at eight knots into Guantánamo Bay and dropped anchors. We hid there for several days, then made a run for New Orleans. We had been aboard for thirteen days by the time we arrived at our destination. Food supplies had run short, and the menu the last four days consisted of dry bologna sandwiches and black coffee.

Dad was a Seabee who landed with the Second Marine Battalion on Saipan, Tinian, and Okinawa, and received a battlefield promotion to chief petty officer. He fought the entire war wearing used, worn-out Marine fatigues, but ironically had to buy a new officer's uniform that was required for his discharge ceremony.

* * *

AFTER THE WAR, we were unsettled for a while, unsure where we wanted to be. Dad bought a new 1948 Ford sedan, which we drove to Florida to see friends. From there we drove in a loop back to Pittsburgh, and then across to Seattle and down to Los Angeles. In LA, my mother saw an advertisement by a contractor offering to build a duplex for about $25,000. She convinced my dad this was too good a deal to pass up.

They signed a contract to build the duplex on a lot in West LA while we lived in a motel and I enrolled at University High School near UCLA and 20th Century Fox Studios. I played guard for the football team and joined the swimming team.

We stayed put for about a year. Dad was hired by the power company as a lineman. He had experience working the "hot stuff," in other words, climbing up into the crisscrossing wires at the top of power poles where he and his partner would deal with fully charged electrical systems. One day he and his partner, a man I knew only as Jack, were working in a maze of wires at a major intersection of power lines. They had placed rubber insulators over the exposed wires and were using "hot sticks," wooden poles that allowed them to move wires around safely in close quarters. Despite the precautions, Jack accidentally came into contact with a high-voltage line.

There was a huge flash as the poor man was electrocuted. He slumped over, hanging upside down in his safety belt. Dad grabbed him and pulled him up tight to his chest and began squeezing, trying to give Jack artificial respiration. For about twenty minutes, my dad supported Jack's full weight while trying to save his partner's life. Finally two more lineman scrambled up the pole, tied lines around Jack, and lowered him to the ground. He had died instantly. My dad's valiant effort was for naught.

When he came home from work that awful day, Dad slumped in a kitchen chair for several hours drinking rum and Coke. I had never seen him so distraught. We shared his sorrow.

"Kathleen," he said. "Let's go back to Panama."

My dad, mother, and Dick soon headed south. I chose to stay in LA to finish high school.

Chapter 4

True Grit and Grace

Los Angeles, California, 1950

Ann Baackes was fifteen, a lovely, vivacious girl with short, reddish-blond hair and a radiant smile. Arriving on her doorstep with a corsage in hand, I politely asked her parents for permission to take their precious child to a school dance. I pulled up in a 1936 Chevrolet convertible, for which I had paid eighty-five dollars. Ann and I both attended University High, though I was three years older.

Ann had been raised by her parents in Westwood Village, a wealthy enclave in West LA, and by her grandmother—heir to a Pittsburgh steel fortune—on an estate in Wisconsin. I never figured out what Ann's family saw in me. They may have liked me because I was polite and a little old-fashioned. Credit my mother, whose training paid off.

One thing I knew for sure: I had found the girl of my dreams.

After graduating from high school in June, 1950, I traveled to Panama to visit my parents. I no sooner got there than my mom said, "Let's move back to LA." So we flew to New Orleans and took a cross-country train. When we pulled into the railroad station in downtown Los Angeles, we spotted Ann jumping up and down and waving her arms. She wrapped her arms around me and didn't let go until she died in my arms sixty years later.

Then, three things happened. I asked Ann to marry me. I bought one-tenth interest in an all-metal, 65-horsepower, wheel-equipped Luscombe 8A airplane. And I enrolled at UCLA, intending to study medicine.

Ann's folks asked us to wait a year, which we were willing to do. I hit the books at UCLA. Flying the little Luscombe was not simple. The plane was a handful with its stiff, narrow landing gear and tail wheel. After an hour of dual instruction, I told my instructor there was no way I would ever learn to fly.

"I can't fly this thing and watch those instruments at the same time," I complained. Remembering this makes me laugh now. All it had was an altimeter, airspeed indicator, tachometer, oil-pressure gauge, and a compass.

Most of my instruction, for which I was paying a dollar and a half an hour, took place after dark. What with attending school, working at a gas station, and romancing Ann, there wasn't enough daylight. One evening after flying up the Pacific Coast to Malibu, where we did stalls and spin recoveries out over the ocean, my instructor told me to return to the airport.

I turned the plane around and headed for Clover Field. I radioed the tower on my single-channel GE radio to advise that I was coming in. They answered with instructions that I was "cleared to enter the pattern for Runway Two Four," which I did, making the required left turns onto a downwind leg, followed by a left turn onto final approach. So far, so good.

As I touched down, however, I noticed the big hangars were on my left side, not on the right, where they were supposed to be. Instantly I realized I was at the wrong airport. I punched the power and began climbing out. I had landed at the Hughes Aircraft field instead of nearby Clover Field. The two fields had parallel runways about three miles apart.

I called the Santa Monica tower again. Those guys apparently knew what I had done. Nonchalantly, I was told, "Yeah, come on over and put it down on our runway." My instructor, who had been a Marine pilot in the war, leaned over and said, "That's why you are called a *student* pilot."

I logged nearly a hundred-fifty hours in this little airplane and another thirty in an Ercoupe. When my instructor decided that it was time for me to make the required cross-country flight, I elected to go to Las Vegas at night. The Luscombe had no electronic navigation gear, so I simply followed the white blinking lights set up on the top of hills to guide the early airmail pilots. For three-hundred miles I listened intently to that 65-horsepower Continental engine for any sign of trouble. The following day, I simply followed the auto traffic heading west on Route 66.

* * *

ONE DAY, a notice on a bulletin board at UCLA caught my eye: "Job offer: interesting work with flexible hours, in Santa Monica, apply below." The prospective employer's name was not mentioned. After applying by mail, I received a letter instructing me to report to an office in downtown LA, where I joined a throng of young men in their early twenties. We sat at desks in a large hall, where we were given a three-hour exam. This appeared to be a high-powered IQ test.

Within a few weeks, I learned that I was among the top sixty scores among several hundred applicants. The first twenty-two applicants to receive top-secret clearances would be offered jobs. I made the cut and received orders to report to a drab, nondescript office building in downtown Santa Monica.

"Welcome to the RAND Corporation," a man said. "Nothing you hear or see in this building can be discussed with anyone. There is a $10,000 fine and ten years in jail if you slip up."

Soon after, an FBI agent walked into one of my classrooms at UCLA, showed his badge to the professor, and asked for me. The professor, believing I was about to be arrested, yelled out harshly, "Sassara, get out of here." Outside, the FBI agent told me, "They need you at work." This was on a Monday. I missed the Wednesday class. When I returned on Friday, all heads turned my way. Just for fun, I announced, "I made bail."

The RAND contract lasted several years. I worked in the company's Systems Research Lab (SRL) trying to figure out how to defend the United States from an attack by the Soviet air force.

* * *

FOUR YEARS after our first date, Ann's father marched this lovely young woman down the aisle and handed her off to me.

Our first problem as newlyweds was dinner. Ann couldn't cook a lick, not even flour, eggs, and water for pancakes. But it didn't matter.

"Don't cry, Ann, I'll show you how," I said.

Unlike Ann, I had the experience. At age thirteen, I had done the shopping and cooked dinners while Dad was in the Navy and Mom worked at Kaufmann's Department Store in Pittsburgh. This prepared me for newlywed cooking. After work, before Ann got home, I often skin-dived around the Santa Monica breakwater and up the Malibu Coast to spear fish and lobsters. One day, she peered into the "icebox," as we called refrigerators then, and suddenly come face to face with the baleful stare of a big sea bass. Ann thought the flesh of all fish was white and came as frozen fish sticks at the grocery store. Whacking off the head and tail of a sea bass was not in her job description.

One afternoon, I arrived home early and fried up a bunch of squid, tentacles and all, which I deviously hid under the mashed potatoes.

"What is this?" Ann asked suspiciously.

"It's a fish called calamari," I told her. "You'll like it."

My lovely wife took a few tentative bites, chewing thoughtfully. Then she noticed a cluster of legs, oddly equipped with suckers on the ends, sticking out from beneath the potatoes. This was no fish. After a closer look, she jumped to her feet and chased me around the house with a broom. She was a joy.

Ann, just nineteen then, worked as secretary to Arthur Dititta, president of Fox Movietone News, a division of 20th Century Fox Studios. Among other things, she greeted and escorted movie stars and celebrities. She was a favorite of Earl Warren, then governor of California and later to become chief justice of the U.S. Supreme Court. Clark Gable sent her flowers. Typical of Ann, she was unfazed by the attention.

We were living in a fancy apartment on Sunset Boulevard, courtesy of Ann's cousin, Bill McGary, who was in the movie business. He had been invited to direct a Broadway play and would be living in Manhattan for several months. While he was away, the apartment was ours.

We loved it. However, there was a little history associated with the apartment that Bill had failed to mention. We discovered he had allowed several young actresses to hide out there while attempting to cool off unwanted suitors. Late one night, Ann and I were sprawled on the bed when someone began knocking loudly at the front door, calling out a woman's name. I slipped on a pair of pants, raced down the spiral staircase, and jerked open the door.

A vaguely familiar older man, about fifty, stood on the step, red-faced with anger and covered with sweat, still creating a racket. "Where's Jean?" he demanded, trying to get around me. This made me angry, too.

"Shut up, and get the hell off my front steps or I'll knock the crap out of you," I told him.

"Where is Jean Peters?" he said. "I know she's hiding here. Let me in!"

"To hell with you," I said. "I'm living here with my wife, so hit the road, or I'll punch your lights out."

"Do you know who I am?" he said. "My name is Howard Hughes, and I want her out of there now."

"Well, my name is Chuck Sassara," I replied hotly, "Remember me when you wake up in the hospital."

With that, I gave him a shove. Howard Hughes stumbled and fell onto his back in the front yard. I slammed the door and locked it.

"Who was that? What was he yelling about?" Ann asked.

"Aw, it was some guy named Hughes looking for his girlfriend—a Romeo and Juliet kind of thing," I said.

Bill McGary confirmed later he had stashed actress Jean Peters in the apartment for a week just before our arrival. I realized I might have responded in a more civilized manner during my encounter with Howard Hughes—a movie-maker, aviator, and aerospace engineer, among other things—who was destined to become one of the world's wealthiest men. I should have invited him in for a drink. Apparently Hughes did catch up with Jean Peters because they were married a few years later.

Speaking of marriages, Ann used to delight in telling people that I could have had Elizabeth Taylor, but chose her instead. At this point in her story, I always kissed my wife and announced that I had made the right choice.

A senior at University High, I was sitting in class one day when an urgent note arrived from the principal, Mr. Wheedon, calling me to his office. He had received a phone call from 20th Century Fox asking for help arranging a date for Elizabeth Taylor, the teenage movie star. Miss Taylor, who had enrolled at the school for her senior year to get away from the studio tutors, wanted to go to the senior prom with a fellow student instead of some young wannabe actor.

"You must have made an impression because she gave your name as her choice," the principal said. "When the studio called, I told them you would be suitable, and that I would tell you what they had in mind."

I didn't have to think about it.

"She seems like a nice girl, but I've already got a date," I informed Mr. Wheedon. "I'm taking my girlfriend, Ann Baackes, to the prom."

Our grandchildren loved this story. Ann never tired of telling it.

Chapter 5
Fritz Runge

Los Angeles, 1954

Fritz Runge graduated from an engineering college in Michigan after World War II, then went to work designing air conditioning systems for General Motors. Fritz was drafted into the Army in 1950 for service in the Korean War. But when he showed up for his pre-induction physical, he asked, "Can I go into the Air Force instead?"

His request was granted and off he want to basic training. When he finished basic, he inquired about the possibility of entering Officers Candidate School. The Air Force gave its approval.

Then, as Fritz neared completion of OCS, he happened to see a notice on a bulletin board offering "interesting work for recent graduates with an engineering background." He applied immediately and was invited to an interview. He understood that if accepted into the program, he would be discharged from the Air Force. This mystified him, of course.

At the interview, Fritz presented his credentials and discussed his recent employment with the Cadillac Division at GM. The interviewer, a fellow engineer, offered him a job.

"Fritz," the engineer said, offering his hand. "I think you and I will make a good team. By the way, forget the fancy titles. Just call me Wernher."

Fritz was the first American employee hired by Wernher von Braun, known as the "Father of Rocket Science," who had designed the V2 combat rocket for Germany during World War II. At the end of the war he and some of his fellow rocket scientists were brought to the United States as part of "Operation Paperclip," a secret government program. Von Braun is most famously known for his work on the Saturn V rocket with a super-booster that propelled the manned Apollo spacecraft to the moon. Fritz became a major participant in the building of von Braun's amazing rockets for the National Aeronautics and Space Administration (NASA).

In 1954, Fritz Runge married Gladys Osborne, a Southern California blonde beauty whose best friend from grade school was my wife Ann. This was how I got to know and became close friends with Fritz. We shared many interests, including airplanes and sailboats. Like me, he loved telling stories.

Fritz told me about his and von Braun's early near-disasters as the two engineers developed enormous rockets.

One of Fritz's favorite stories was about setting out to launch a rocket that would not ignite. It was standing on a launch pad, switches in the "go" position, ready to fire. But when the start button was pressed, nothing happened.

Von Braun, Fritz, and others in the launch crew were positioned behind a blast fence several hundred feet back. A coaxial cable had been laid in a ditch out to the pad, where one end had been attached to the bottom of the rocket. Von Braun looked around for a volunteer to crawl in the ditch out to the rocket to inspect the connection. Without a word spoken between them, Fritz knew that as von Braun's most trusted colleague, he was the one to go.

Fritz told me he was just reaching up under the rocket to unwind a coupling when von Braun peered out an observation window with his hands over his ears. Apparently von Braun was renowned for his humor.

Later Fritz helped design the International Space Station. One of my fondest memories was of him and Ann down on his living-room floor studying huge photos of the station as Fritz patiently explained how it was put together and why it was needed for the advance of science.

Chapter 6
North to Alaska

Los Angeles, 1955

While studying at UCLA and working for the RAND Corporation, I met a fellow researcher who had been to Alaska. He told wonderful stories about the natural beauty of the mountains, glaciers, forests, rivers, and lakes; the abundance of fish and game; and the freedom and independence that Alaskans experienced in the vast, roadless territory where airplanes were the main form of transportation.

The more he talked about Alaska, the more intrigued I became. Naturally I wanted to take a look for myself. Frankly, I didn't care much for the brown, parched terrain of Southern California. Because my family had always moved to where the work was, moving was no big deal.

"Let's go up north and see what it's like before we settle down," I said to Ann. She was easy to persuade. I would graduate from college in two weeks. I quit my RAND job and Ann submitted her resignation, too. We began packing.

On the big day, we jumped into our 1954 Volkswagen bus and drove north, followed by my parents in a pickup truck on which Dad had mounted his homemade camper. Everyone in my family had the travel bug—"itchy feet" we called it—so off we went without a lot of fanfare.

The rough road began in British Columbia, where the Canadians were carving out a new highway through the Fraser River Gorge. Farther north, we felt like pioneers driving the remote Hart Highway to connect with the Alaska-Canada Highway (the "Alcan") at Dawson Creek, B.C. For most of the next 1,400 miles, we encountered an abundance of mud, rocks, and thick dust; changed countless flat tires; and replaced a broken axle near Whitehorse, Yukon Territory.

At the end of this two week ordeal, we rolled into Eagle River on the outskirts of Anchorage and set up camp.

On our way into the city the next morning, we passed what appeared to be at least a thousand light planes tied down on Merrill Field. I had never seen such a busy airport devoted exclusively to light aircraft.

"Look," I marveled, "there must be eight planes in the pattern."

We also visited Lake Hood, where a control tower directed floatplane traffic. Next, we drove to nearby Anchorage International Airport, where a gaggle of big transports mixed in with several hundred more light planes on wheels. Parking in front of the small terminal building, I walked into the first office I found. The sign said, "Pacific Northern Airlines." This sounded promising, so I asked to see the boss and was introduced to Roy Buckles, the airline's station manager.

"Mr. Buckles," I said, "I just graduated from college, I love airplanes, and I even know how to fly. I'm ready to go to work."

I was hired on the spot and told to report for work at six a.m. the next day. As we left the airport, I noticed a sign warning: *Aircraft have the right of way on roads.*

I knew right then Alaska was the place for me.

Pacific Northern Airlines operated Lockheed Constellations, one of the world's most beautiful aircraft. A collaboration of science and art. A long lean machine with a distinctive curved shape and triple tails.

Our direct competition, Northwest Airlines, flew Douglas DC-7s, a great airplane as well. The DC-7 had all the sex appeal of a dump truck. Strictly utilitarian.

I started my work day at the new Anchorage terminal at 6 am. I would help out at the counter, writing tickets and checking in passengers for the first flight of the day, Anchorage to Seattle, Washington, five hours and twenty minutes away.

The whole gang – ground crew, ticket agents, dispatchers and pilots – pushed and pushed to make sure we were number one for takeoff at 8 am. – the exact same time Northwest was scheduled to leave.

Both of these big four engine planes flew at the same speed, so the race was on. Every known trick was used to squeeze another knot out of them - slight changes in altitude and odd trim settings such as slightly nose up with the plane wanting to climb ever so slightly.

The captain and first officer took turns pushing forward on the control wheel in an effort to stop the oscillation of the plane known as fugoids. This yielded maybe one or two knots.

When they landed in Seattle we would get a telex telling us who had won. Great fun!

My other job was doing the weight and balance for the flight. I would assemble all of the figures such as fuel on board, cargo weight, baggage weight and number of passengers. All this was put on a slide rule and it was slip sticked to come up with a weight and balance. When I was done calculating this I handed it to the dispatcher. Sometimes I handed it directly to the pilot. The time spent in the dispatcher's office put me in direct contact with the flight crew.

One girl I met, a so-called beanie, or flight attendant, was Ellen Wilson, a happy blonde from Opportunity, Washington. I kept running into her in the back of the cabin after a flight up from the lower 48 states.

As soon as all the passengers got off the plane she would rush back into the galley and fill bags full of uneaten snacks – mostly apple pandowdies. "Hey Ellen, gimme some of those things" I'd say. "Sure," she would answer. "They would only go to waste anyway."

I told her my younger brother Dick was on his way up to Alaska to get sworn into the Army. He had just gotten drafted and I had arranged for him to report in Anchorage rather than Fort Ord, California.

"You've got to meet him. He's a good guy and full of adventure – just your type."

Sure enough, they met, dated, married and raised two great kids.

The gang at Pacific Northern Airlines was like a family. Everybody knew everyone who worked for them throughout their system starting with Art Woodley, president. Art had started with one plane in Alaska. I think he was a bit of a scoundrel, at least according to Bob Reeve who also started with one plane flying to mining camps up in the Wrangell-St. Elias Range.

In the book about Reeve, *Glacier Pilot*, there is a picture of Reeve's shack on the airstrip in Valdez with a sign on it that announced "Stay Away From My Tools, Especially You Art Woodley."

Pacific Northern Airlines merged into Western Airlines which later merged with Delta Airlines. So it goes in the airline business – feast or famine. It's no pun to say the airline business has its ups and downs. I worked for PNA for about 10 months where I got to learn about weight and balance on large aircraft. This really paid off when in later years I was flying in and out of third world countries where ground crews just stacked cargo willy nilly inside those huge fuselages and just walked away! One time I was in Honduras picking up a load of fresh beef when I noticed the nose wheel was mashed flat and the strut was fully compressed! The tail of the Constellation was up in the air. A quick calculation on my slide rule indicated we were over 20,000 pounds too heavy in the forward section of the plane! A disaster averted – thank you PNA.

During the ten months that I was employed by PNA, Ann and I lived in a little house on Nugget Avenue. It was right behind the joint called Chilkoot Charlie's. It's still selling booze at that location in the year 2014. Ann was pregnant with our first son. She was learning how to go shopping at the Piggly Wiggly grocery store while wearing snowshoes. A big adjustment from the bright lights of Hollywood.

My dad was working for the C.A.A. (Civil Aeronautics Administration) at various locations around the state and my mother was holding down the fort out at our new-to-us lodge on Big Lake.

FIFTEEN YEARS LATER

I was invited down to Los Angeles to meet with some of the titans of industry during Alaska's push to get approval for the construction of the 800 mile pipeline from Prudhoe Bay to Valdez. Almost all of the right-of-way would be on federal land. A resolution by the United States Congress was required to get a lease.

Some of Alaska's leaders were dispatched to garner support for its passage. I was majority leader in the Alaska Legislature.

I drew the California Chamber of Commerce meeting which was being held at the Century City Plaza. It sat right on top of the spot where my wife Ann had been when she worked for 20th Century Fox Studios.

I shared the head table with then Gov. Ronald Reagan and Terry Drinkwater, president of Western Airlines. Reagan introduced me, making it sound like I was a backwoods rube from Alaska. Hardly even housebroken.

I thanked him and went into an "aw shucks" act, sort of alluding to the pleasures of civilization, like for instance, flush toilets.

Then I said in a very clear voice, "Yeah, boy I think I was standing on this spot in 1951 when I proposed to my wife Ann, who at the time was secretary to the president of Twentieth Century Fox Movie Tone News.

I glanced over at Terry Drinkwater and winked. He was enjoying the repartee. Reagan was staring at the floor. I think he was pissed off at me. Later that day I went over to him and stuck my hand out.

"I met you about 10 years ago when you called University High School and asked the principal if he knew a couple of boys who would like to earn a dollar an hour. You and I and some other fellows set up chairs, hundreds of them in the Hollywood Bowl. You were president of the Screen Actors Guild then. There must've been some sort of concert or fundraiser. Anyway, you and I both earned our keep that day. I know I was plenty hot and sweaty."

Reagan took another look at me, trying to place me. Well anyway, we were off on better footing. I was often in Washington D.C. for the next few years working on issues that affected Alaska. He was always cordial to me.

When the speeches were over, business cards exchanged, and promises made, Terry Drinkwater grabbed me and pulled me aside to a smaller room where he and his buddies were going to have a drink and relax after a hard day of politicking.

As we settled in, me with my soda pop, Terry asked me about the flying activity in Bush Alaska. I began filling him with stories about flying floats and skis. He really wanted to hear about the de Havilland Beavers. And the Pilatus Porters. I've never met an airline pilot who didn't want to get his hands on a real Bush plane. By the time you get up into big jets, the fun is gone.

I told him about a recent flight. I had met a senior captain with British Airways named Smythe. I told Capt. Smythe to bring some old clothes with him the next time he stopped in Anchorage. We arranged a time for me to pick him up at the Captain Cook Hotel.

I took Capt. Smythe out to Merrill Field where I had a rather beat up old Cessna 185. We threw some sack lunches, shovels and buckets on board. An hour later we landed on the exposed tide flats. There were already 15 or 20 planes sitting on the sand. At high tide there would be nearly 30 feet of water where the airplane was sitting.

He looked around in amazement at all the activity. There were planes landing and taking off all over the place. "What's going on here?" he asked. I said, "Grab a shovel and follow me." We walked about 15 feet, dropped the buckets and began digging clams. Capt. Smythe said "they'll never believe it." "Never believe what?" I asked. "The guys back in England in my flying club. We had four Piper Cherokees that we take out on weekends or days off."

"We fly from airport to airport having lunch, occasionally going over to France. It is all very nice but this is wild. I've never seen light planes being used like this. This is great!"

In 45 minutes we had dug up two five gallon buckets full of big razor clams.

I took him to my home where more clam diggers had gathered. We fried clams and drank beer for hours. Capt. Smythe went around and shook hands with everyone. He said he had never had so much fun in an airplane. I told him "Next time you come over we will go shoot a moose."

When I told Terry Drinkwater this story he started jumping up and down and shouted, "Let's go get some clams."

I really like the guy; he was real. I asked how he got started.

"Well," he said, "It was like this! I was driving a Model A Ford up and down the length of the California trying to raise money for more airplanes. I got up in the morning, got out my scissors and some cardboard and made new insoles for my shoes – then hit the road. Nowadays I've got a bunch of big Boeing jets and I get up in the morning, pick up my scissors and cardboard to make new insoles for my shoes and hit the road! That's the flying racket."

* * *

AS ANN AND I settled into a log cabin in Anchorage, we looked around in delight and wonder. It was so different from our life in California.

We were overwhelmed by the frontier atmosphere and the beauty of the mountains, trees, and intense green foliage. Quickly I came to understand why so many people drawn to Alaska for the gold, oil, and fat paychecks had chosen to stay permanently for the quality of life.

On my days off from Pacific Northern Airlines, we drove the VW bus throughout Southcentral Alaska in ever-widening circles looking for the perfect place to live. Visions of hunting and fishing lodges with a plane parked nearby popped into my head as we drove north on the narrow, bumpy Glenn Highway to Sheep Mountain Lodge, which was for sale. The property had its own airstrip, and white Dall sheep dotted the steep mountainside behind the lodge. I visualized selling steak dinners and sourdough pancakes to tourists and guiding hunters into the surrounding mountains.

We also drove south on the unpaved Seward and Sterling highways to the Kenai Peninsula, where we were delighted to find the Russian River, one of the world's best fishing spots. Having grown up skin diving and spearing fish, I immediately stripped off my shirt and pants, plunged into the cold river, and began grabbing fish and flipping them onto the shore. Ann beheaded them. After cooking and eating a couple of fish on the spot, we stuffed the rest into a pillow case for the trip back to Anchorage where we had a big fish fry at our cabin on Fireweed Lane.

* * *

ARMY CAPT. Kenneth Lilly, whom we had met coming north on the Alcan, stopped by the cabin one day and entered without knocking. "This is Klondike," he said, dropping a forlorn, bedraggled husky puppy on the kitchen floor. Our friend retreated hastily in his Pontiac station wagon without another word.

We put the puppy into a cardboard box the first night, unsure what to do with him. Our plans did not include hauling around a dog. The whimpering and wailing began as soon as Ann and I went to bed. Klondike was unhappy. We tried to ignore him.

"Did you hear that?" Ann asked. "I swear he called out mama."

I had heard it, too. So I brought the box into the bedroom, depositing it at the foot of our bed. All was quiet for the rest of the night. Klondike slept at the foot of our bed for the next sixteen years without complaint.

* * *

By then, the Sassara clan had been in Alaska for two months and it was time to make a decision whether to move south to Kenai or Homer to be near saltwater, or move eighty miles north of Anchorage to Big Lake, where we had found a small lodge for sale on six acres with 3,000 feet of water frontage.

The lodge was a modest log structure—rough and unfinished on the outside, but nice inside. It had running water, though the toilets were outside. The lodge had a great bar and dining room with a full restaurant kitchen, including a large stove with a big grill surface and double ovens.

Ann and I and my parents met with Bob and Katie Payton, who wanted to retire. They were offering the property for $25,000, including a liquor license issued by the Territory of Alaska, four or five twelve-foot skiffs, and a bunch of outboard motors. What a deal!

The four of us looked at one another and began nodding in unison. "We'll take it," my mother said.

We shook hands all around. This handshake commitment was solid gold, Alaska-style. A week later, we signed the papers and handed over a check. The terms were a down payment of $5,000 against the balance to be paid off in $5,000 annual payments over the next four years. This was a lot of money in 1955, but we never regretted making the deal.

My mother mostly worked alone keeping the lodge open while Ann and I lived in Anchorage and Dad often traveled for work. I continued to work at PNA while my dad signed on as an electrician at the Civil Aeronautics Authority (later to become the Federal Aviation Administration in 1958), which dispatched him periodically to the far corners of Alaska to wire runway lights and emergency generators.

When the snowfall at Big Lake set a record in 1955-56, we understood why the previous owners always closed for winter.

On my days off Ann and I would head north to Big Lake to lend a hand at the lodge. I was learning to drive on ice and snow. The novelty almost wore off during one trip when we got stuck in the middle of the highway in a blizzard west of Palmer. We were driving Pappy's 1947 two-wheel-drive International pickup with chains and a shovel. It took us nearly four hours to move one mile, shoveling our way in front of the wheels, then jumping in and ramming it forward until we spun out again. Finally we made it to Koslosky's Store in downtown Palmer. We found a little hotel nearby.

"Ann, I can't do this anymore," I said. "I'm going to call PNA and tell them I'm going to be a day late for work."

We had discovered the true meaning of winter.

* * *

ONE MORNING my mother was sipping her first cup of coffee when she spotted a moose standing in belly-deep snow in the parking lot about twenty-five feet from the back door.

"Pappy, Pappy," she shouted for my father, "get a gun and shoot it!"

My dad, still half asleep, grabbed a 30.06 and eased the door open. Without much thought he took aim and pulled the trigger. Down went the moose. Then Dad put on some clothes and waded out into the snow to make sure it was dead. The moose had no antlers, so he couldn't see whether it was a cow or bull. But it didn't matter. The hunting season was closed, and a cow moose could never be taken in any case.

After a quick consultation, my parents set to the task of butchering what they figured to be a thousand-pound animal. They worked all day skinning it out, cutting it into various cuts that they dragged into the lodge and up the stairs to hang in the attic. Using birch poles, they pushed the legs and guts into a fishing hole in the thirty-inch-thick ice.

The head and hide were not disposed of so easily. Mom and Dad had to widen the fishing hole to get rid of the head. The hide would not stay submerged, no matter what they tried, so they pulled it out of the water and dragged it over to the lodge and hid it in a crawlspace.

As darkness descended, the scene was a bloody mess. While my mother pointed a flashlight, Pappy shoveled fresh snow over the blood, and then covered the worst of it with a couple of white bed sheets.

They were worn out, yet satisfied as they crawled into bed, secure in the knowledge they hadn't been caught taking an illegal moose.

When daylight returned, my parents discovered that foxes and ermine, with help from a squadron of ravens, had dug into the bloody snow scavenging bits of moose flesh and blood. They cleaned up the mess again and fired a few rounds from a shotgun to scare away the ravens. This scene repeated itself for four days.

The next weekend, I chained up the VW bus and we drove to Big Lake by way of Palmer and Wasilla. As usual, it was rough going with lots of rocks and potholes. The major industry in this area seemed to be repairing flat tires. We got to within a mile and a half of the lodge, then skied in the rest of the way.

Ann and I were proud as punch, having brought food supplies, including a smoked ham. Mom and Dad pawed through the groceries, laughing as they pointed to the stairs leading up into the attic. Up we went. We could not believe our eyes. There was meat everywhere.

"This will get us through the winter," Ann observed.

Chapter 7
The Yacht Club

Big Lake, 1956

After about a year in Anchorage, Ann and I moved to Big Lake, where we assumed a more active role helping operate the lodge our family had christened the "Yacht Club." The timing was good. Big Lake was a rough eighty-mile drive from Anchorage, but it was only twenty-five air miles across Turnagain Arm, and had become a magnet for a growing number of fledgling pilots looking to try their new wings away from the close supervision and confinement of the city.

Big Lake was bustling with boats and airplanes day and night. Two regulars at the lodge were Herbert Mensing and Jack Peck, who loved my mother's cooking. They owned Alaska Aeronautical Industries, the Cessna dealership based at Lake Hood. I looked forward to their visits.

The two men knew I loved airplanes and always looked for me when they arrived. "Hey, Chuck," Herb would say, "quit washing dishes for a few minutes and go see what we're flying today."

They didn't have to ask twice. I pulled off my apron and rushed out to the airstrip or down to the dock. These visits gave me an opportunity to crawl over, and occasionally fly, most of the two-place airplanes available in those days—Taylorcraft, Piper J3 Cubs, Champs, Chiefs, Aircoupes, and Cessna 120s and 140s. One time, Herb and Jack loaned me a Piper PA-16 Clipper on floats with a 108-horsepower engine.

It took some doing to get the Clipper into the air loaded with the two heavy men and their hunting gear. Having short wings and no flaps without a lot of power, it had to be nursed up onto the step, then run quite a way down the lake to get off. When I bought a Taylorcraft on floats with an engine that supposedly put out 65-horsepower, I was surprised to discover it handled like a Piper PA-18 Super Cub. A few hours in that Taylorcraft with its long wings and superior airfoil showed me its impressive capabilities. It couldn't beat a Cub off the water, but the Taylorcraft would win a fly-off against everything else, including a Cessna 180.

One day, big Bill Borland, chief pilot for Reeve Aleutian Airways, wrapped an arm around me and said, "Come on, I'll show you how to stay alive." Every chance we got, we would load the Taylorcraft with enough food for a small army. Bill loved to eat. And away we went.

First time we took off, I pulled up rather abruptly, nose high. Bill reached over and shut off the engine. Down we went, flat onto the water, scaring the hell out of me. I looked over at Bill, who asked, "Did you learn anything?"

"Yeah," I answered. "*Fly* it off."

We had hit the water so hard that the engine ran rough when I re-started it. I shut it down, crawled out onto a float, and raised the cowling. Three of the bottom sparkplug wires had come off. They were the old-fashioned, unshielded kind that just clipped onto the top of the plugs. I turned around to tell Bill what had happened and there he sat with a big ham sandwich stuck in his face, his eyes crinkled up in laughter. Lesson learned. For the next 25,000 hours I made certain to fly airplanes with respect.

Bill's father, Earl, had been an aircraft mechanic in the 1920s. He was killed in a 1929 crash with pilot Carl Ben Eielson, who was attempting to rescue passengers stranded on a ship frozen off the Siberian coast with more than a million dollars in furs aboard.

Bill grew up in Fairbanks, learned to fly at age fifteen, and joined the service in World War II. He went on to get his wings and a quick ticket out to the Pacific where, at eighteen, he became one of the youngest B-26 bomber pilots in the war. When Bill got home, Bob Reeve made him chief pilot of his airline, flying Douglas DC-3s and the four-engine DC-4s.

Reeve Aleutian Airways served the Alaska Peninsula and Aleutian Islands, the latter a region known for having the worst weather in the world. Staying alive was no easy trick. Bill and his pilots flew over mountains and lots of cold water while encountering icing conditions, snow, fog, and rain. The wind came from every direction. There was no room for error.

"Chuck, I hold the unchallenged world record for 180-degree turns," Bill said. "I want you in second place."

About this time, I bought a small portable radio, an Air Boy Senior with a dry cell battery, tunable receiver, and two crystals dangling at the end of a string attached to the tuning knob. You just picked up the crystal, looked at the number stamped on the back, and plugged it in.

My new radio got a test when I flew from Big Lake to Galena to visit my dad, who was installing standby generators at the airport. I had used my paper charts and eyeballs on the way up there, then, as I got close, put on a headset and followed the four-course radio range into the station at Galena. The results were impressive. I stayed on the steady tone coming through the headset and listened as the signal got louder, then went silent. I did a 180-degree turn and listened intently to pick up the signal again. There it was, a steady tone getting louder. Wonder of wonders, the FAA antenna was straight ahead.

One day I arrived back at Big Lake in a heavy snowstorm. I landed next to the Yacht Club just as four guys headed my way. They were crew members of an Air France jetliner that had just flown over the North Pole from Paris to Anchorage, one of the first such flights. We shook hands while they inspected the Taylorcraft. The visibility was about half a mile with an indefinite ceiling. Curious how I had found Big Lake in the storm, they wanted to see my instrument panel and radios.

First I showed them how I kept track of fuel. "I just follow the bent wires sticking out of the nose tank," I said, pointing to a wire attached to a cork inside the tank. It was the world's least-complicated fuel gauge.

Then I showed them the compass and a cheap pocket watch hanging on a string attached to the windshield brace. "That's my turn and bank," I said, tongue in cheek, "as long as it hangs straight down, all is well. When it points straight up, you have a problem."

The Frenchmen laughed like hell at this nonsense. But they got serious when I showed them the stack of radios—my portable Air Boy Senior with both crystals dangling from the tuning knob, accompanied by my homemade direction-finder. I had gone to a five-and-dime store where I tried out every small portable radio in stock. I put them on the counter, turned them on, and tuned each one to a broadcast station. Then I would rotate them back and forth and listen to hear if the volume would go up and down. I was listening for a null—a drop in volume. Some of those cheap radios had a loop antenna inside instead of an external whip.

The way it worked was simplicity itself. You turned on the radio, placed it on top of the panel, found a broadcast station, and turned the volume down low and listened as you moved the radio left, then right. When the signal faded, you knew the station was ninety degrees to the back of the radio. Presto, you had an automatic direction-finder for about eleven dollars and fifty cents, batteries included. By this time, the Air France crew was howling and slapping me on the back. Just as we began to walk back to the lodge, we heard a plane coming in. We stopped talking and listened. It sounded like a Cessna 180.

Sure enough, it was a 180 on skis coming in for a landing on the frozen lake. However, the pilot was landing crossways to a road that had been plowed on the ice. The snow berms created on either side by the plow were about two feet high and as solid as concrete. The pilot made a perfect three-point landing about twenty feet short of the snow berm. When the skis buried themselves into the berm, the Cessna came to a dead stop. It split open like a cantaloupe, both of the doorposts shearing off at the top near the leading edge of the wings. The floor folded over at about ninety degrees. The pilot and his stunned passenger sat there strapped into their seats. The engine was nose down in the snow. We jogged over to the wreck to check on the condition of the two men, who were cursing a blue streak.

The men unhooked their seatbelts, planted a foot in the middle of the instrument panel, and stepped straight ahead out of the plane.

"Is this the usual thing seen around here?" one of the French pilots asked.

"No," I said, "usually they flop on their back and slide farther."

* * *

ONE WINTER DAY, a De Havilland Beaver on skis parked in front of the lodge, delivering three Air Force officers who had come for lunch. While they waited for hamburgers and fries, I entertained them with flying tales when an Aeronca Champion on wheels dropped down low to look at roads plowed on the ice that doubled as landing strips. Three ice roads heading out across Big Lake from the lodge like spokes on a wheel.

The pilot dropped down to about fifty feet, checked out the lake surface, and went around to set up for a landing. I noticed he chose to land crossway to the wind. Here we go again.

"Hey, you guys, grab your coffee and let's go outside and watch this guy crash."

The Air Force pilots, Ann, my dad, and I lined up in front of the lodge watching as the Champ made a final approach. It was sort of like shooting a movie—we knew the script. The road had been plowed by a snow plow with a six-foot blade. Three passes had produced a passable runway fifteen feet wide with a two-foot berm on either side.

The plane was only about twenty feet off the ground when it glided past us. The pilot set the plane down perfectly in the center of the road on three wheels. Then the cross wind went to work, pushing the aircraft sideways. In an instant, the right main wheel hooked into the berm. The plane did a turn like a ballerina and flipped over onto its back.

We walked over to the plane, still clutching our coffee mugs, and peered into the cabin. A torrent of profanities assaulted our ears. A passenger was crawling out, angry as hell.

"I told you not to land, you dumb bastard," the passenger said.

We walked back to the lodge to eat lunch, satisfied the entertainment was over. But no, there was more. It turned out the plane had been rented from Bill Barton at Merrill Field. When I found this out, I showed the shame-faced pilot our newly installed telephone. But he wanted a stiff drink. An hour went by.

"Hey, you better call Barton," I said. "He will be worried sick."

Another hour went by; still no call. I finally got the story from the passenger. Seems he was offered a ride by his buddy, the pilot. They went to Merrill Field to rent the Champ for an hour of sightseeing. Bill Barton told them not to fly to Big Lake with his plane. This was the passenger's first ride in a light plane, and he had no idea where he was until he crawled out of the wreck, stood up, and looked around.

"Where am I?" he asked.

"Big Lake," I said.

So went the humdrum life of a lodge operator in Alaska in the 1950s.

* * *

WE HAD INSTALLED a seven-hundred-fifty-gallon tank and fuel pump on our dock. We sold only eighty-octane gasoline for airplanes and outboard motors.

One night a big old Cabin Waco on floats pulled up to the dock for fuel. Ann went out to take care of them. She turned on the pump, pulled out about fifty feet of hose, and handed it to the pilot who was clambering up the side of the plane. Three passengers wandered over to the lodge for a quick bite.

Ann sold the pilot about forty gallons. Then he and his passengers climbed back in and taxied out into open water for takeoff. The pilot applied full power to his old, round Jacobs engine, straining to get up on the step with the nose high and water blowing everywhere. No good. So they pulled the engine back to an idle and circled around while letting it cool down.

Another try, using the same technique. Elevator pulled all the way back. It still wouldn't climb up on the step. So they shut it down again and circled around for yet another try. This time the pilot tried to rock the Waco up on the step while crossing over the wake he had created on his first try. I was inside the lodge listening to this when suddenly the sound of the engine changed. I looked out the window just as the Waco came into view. It was up on the step with the engine snarling and spray flying. Thank God, I thought, they were underway at last.

I watched as they ran down the lake about two and a half miles before going around a point and out of sight, still on the step. Behind the point, the lake extended for another two miles. According to several witnesses, the plane finally broke water with about four-hundred feet left.

A row of swamp spruce lined the shoreline dead ahead. The pilot, with the plane barely flying, pulled back on the wheel trying desperately to clear the trees. That was it for the old Waco. The plane stalled and fell about fifty feet, smashing into the lake nose first and starting to sink.

The pilot and passengers were beat up, shaken, and bleeding. All four swam to a dock about a hundred feet away.

A man was sitting on the dock, feet dangling in the water. He had a glass of whiskey in one hand and a toothbrush in the other. He had been listening to the sound of the engine as the Waco approached and had a front-row seat when the plane did its last swan dive.

Coincidentally, the man on the dock was Al Hulen, head of the FAA in Alaska. The pilot did not possess a seaplane rating. He lost his license to fly.

Chapter 8
The Aeronca Sedan

Big Lake, 1957

I flopped onto a couch in our lodge at the Yacht Club, weary at the end of a tough day at work, followed by the long drive home from Anchorage.

"How was your day?" my mother asked, handing me a cup of coffee.

Mom ran the lodge and sometimes found herself stuck there alone, especially off-season. My Dad was away on an FAA assignment. I had found a new job selling radio and television commercials for KENI AM and TV. Ann had gone to LA to visit her family, taking our son Charlie with her.

When Mom got bored, she would put on the damnedest get-ups, mostly surplus winter gear from the Korean War, including white rubber inflatable boots and an ankle-length wool overcoat. All she needed was a steel helmet and she could have joined the cast of the TV comedy, *MASH*. She had cut a permanent hole in the lake ice, which was about four feet thick. She fished for hours, jigging a lure into the hole and pulling out lake trout and ling cod. She kept the hole from freezing over by placing a sheet of plywood over it.

"Funny thing happened today," she said. "I bought a plane."

"You did WHAT?" I jumped to my feet. My mother was grinning like a mad woman. She knew this news would excite me, and it did. I could hardly contain myself.

"Tell me, tell me!" I shouted. "Are you kidding?"

"Well," she said, relishing her story. "Jack Peck and Herb Mensing flew in this morning with an airplane they wanted to show off. It was in good condition, they said, and would be perfect for us."

"What kind of plane is it?" I asked.

"I don't know," Mom replied, "but it's big and roomy and looks beautiful, so I wrote a check and signed on the dotted line. It's ours now."

I was overwhelmed. What a mother I had.

"Jack and Herb said they would have it ready for you to pick up in the morning at Lake Hood."

I didn't sleep much that night. By dawn I was on the road heading for Lake Hood. My heart pounded as I pulled up to Alaska Aeronautical Industries, looking around for an airplane fitting my mother's description.

There it was in the hangar, the most beautiful plane I'd ever laid eyes on, a red and white Aeronca 15AC Sedan glistening under the lights. The interior was gorgeous. Everything was new or updated, including an over-hauled engine and a new propeller, tires, brakes, windshield and windows, even fabric on the fuselage. It was basically a new airplane.

Jack and Herb enjoyed my delight. We pushed the Aeronca out of the hangar and I jumped in. This four-place plane was *big*. Until then I had flown only two-seaters, except for a Piper P-16 Clipper on floats. This powerful airplane was a big upgrade for me.

I fired up the Aeronca and waved goodbye. Then I applied power and kicked the rudder a bit to break the skis loose and slide down the ramp onto the frozen surface of Lake Hood.

The tower, which had been watching this activity, flashed a green light clearing me to taxi. Like most light planes in the 1950s, the Aeronca did not have a radio. I taxied down to the west end of the lake, then turned around. The tower again flashed me a green light. Time to go.

I opened the throttle, marveling at the smooth power of the six-cyl-inder, 145-horsepower engine. I was madly in love. As the Aeronca raced across the bare ice, the tail lifted ever so slightly and we were airborne. Climbing to nine-hundred-fifty feet, I joined the traffic pattern for Merrill Field at the other side of town. I was third or fourth in line to land.

At the time, Merrill Field was the fourth-busiest general-aviation airport in the United States. It logged about 235,000 operations annually, mostly planes relying on light guns from the tower for directions.

I slid to a stop in front of the FAA office. My mother had financed the airplane, so we needed insurance and I had to add a land-plane en-dorsement to my pilot's license. I had gotten my private pilot's license for floatplanes in a two-place Taylorcraft 12C, but had never gotten around to getting a land-plane endorsement, although by then I had flown for several years in various aircraft, both wheeled and on floats. Mr. Gruber of the FAA checked my paperwork as he prepared to give me the required check ride.

"Now, you're not going to tell me that you never carried a passenger in all that time, are you?" he asked. My license allowed me to fly passengers on floats, but not wheels. Before I could think of what to say, he said, "Come on, let's go out there and see what you can do."

After the preflight inspections, we climbed in, and I fired up my new best friend.

"How much time do you have in this airplane?" he asked.

"Six or seven minutes," I said. Merrill Field is only three miles from Lake Hood.

Mr. Gruber put me through the paces—short field, rough field, and wheel landings. He had me make a three-point landing on a designated spot while pretending to talk with the tower on an imaginary radio. Satisfied, he issued me a new license for "airplane single-engine land and sea" up to 600 horsepower.

As I left the FAA office, Mr. Gruber reminded me I was little more than a student pilot. "Don't do anything stupid," he admonished me.

For the next three years, I practically lived in the Aeronca, polishing my skills while flying throughout Alaska and earning a few bucks here and there. United Geophysical out of Calgary was looking for oil in the Matanuska-Susitna Valley. I was asked to haul drill bits to remote sites. Instead of invoicing the company for flight time, not having a commercial license, I billed the flights as cabin rentals at the Yacht Club.

A year after Mr. Gruber turned me loose, I was cruising at about ninety miles an hour at 4,000 feet, having descended from 7,500 feet on my way to Merrill Field from the village of Galena on the Yukon River. Suddenly an Air Force jet fighter streaked *under* the Aeronca and pulled almost straight up, scaring the hell out of me.

Back at Merrill, I walked across the street to Peggy's Airport Cafe for sourdough pancakes, proceeding to tell my story about the close encounter. Peggy's was a popular gathering place for the hundreds of pilots who flew out of Merrill Field. Naturally everyone had an opinion. The consensus of the coffee-drinking pilots was that I had the right-of-way because I was at the proper altitude and had been overtaken from behind.

I was stuffing my face and still talking excitedly when in walked Mr. Gruber.

"Tell me what happened out there," he said. "The Air Force claims you caused a near-miss."

"I don't think so," I said.

I put down my fork long enough to recite my version of the encounter, including point of departure, course, and altitude. This information was vitally important. In the Anchorage area, traffic is so dense with overlapping traffic corridors and complex flight patterns that no one can afford to be off course or outside of one's assigned altitude.

Near-miss reports would have to be submitted. The U.S. Air Force already had filed its report, probably figuring the best defense was an offense. Why not blame the civilian pilot? Back at the FAA office, Mr. Gruber looked into the Aeronca and checked my license. The ink was practically still wet.

"Chuck," he said, "don't worry about this. There is no doubt you had the right-of-way. I already looked at the radar track. You're in the clear."

The FAA inspector had chosen not to ask me about the pile of drilling bits tied down in the back of the Aeronca. This was not a legal load for a private pilot without a license to carry freight. He had done me a favor, and I never forgot it.

* * *

THE AERONCA SEDAN was good to me. In summer, it was tied down at the Big Lake airstrip. In winter, it sat on skis about fifty feet from the lodge, ready to go. It was in the air almost every day—flying to Palmer for groceries, to Wasilla for mail, to distant hunting grounds for moose and caribou. We hauled sick and injured folks into Anchorage and newborn children back.

Our second son, Richard, was born in January, 1959, the same month Alaska became a state. Richard was three days old when he made his first flight.

I learned a lot about the basics of flying including airmanship, fuel management, navigation, and pinpoint dead-reckoning navigation. Dead-reckoning is flying without radio aids, taking into account factors such as compass course, wind direction and velocity, and location of known rivers and mountains. Because wind direction and speed are so important, experienced pilots read clues in nature as they go. Clue number one: the shiny side of a lake surface indicates the side the wind is coming *from*. Clue number two: streaks of foam on the surface of the water suggest the wind velocity is at least ten knots. Clue number three: the behavior of leaves on the trees can tell you wind direction.

Few airstrips were available in those days. The Aeronca carried only thirty-six gallons of fuel, so I always carried extra gas in five-gallon cans. I was always on the lookout for a place to land for refueling.

The Aeronca didn't have flaps but she was a sure-footed companion. I became adept at putting her down exactly on a spot—not *close*, but *right on the spot*. Often I would tie an old shirttail to some brush to serve as a wind sock, which I watched closely prior to opening the throttle for takeoff.

I spent many a night out alone sleeping under a wing, looking at stars and listening to wolves. She took me up into the Arctic and as far away as the Aleutian Islands. The only time she failed me was over water on an approach to Anchorage. I was at about 4,000 feet, aiming for the runway at Anchorage International Airport, when the engine began spitting and sputtering. I had no radio to contact the tower. I wasn't sure if I would make it. But I was lucky. I landed on Runway Six, rolled about one-hundred-fifty feet, and then the engine quit altogether. The tower sent out a tow truck. Of all things, the culprit turned out to be fuzz from a chamois skin that had collected on the carburetor, choking off the fuel flow. In those days, we always filtered our fuel through a chamois to catch dirt and water collected in the bottom of the can.

I put about six-hundred hours on the Aeronca before selling it to Fred Notti. Next came a Bonanza 35, a fast airplane that was relatively inexpensive to fly. I learned to love this plane. Yet, it was instantly for sale, just like the other two-hundred or so aircraft I've owned. This was the nature of my business: I wasn't going to make any money if I didn't sell airplanes. In a few months, I sold the Bonanza and bought a Cessna 180 from Jay Hammond, a member of the Alaska Legislature who later was elected governor. Those early 180s were fantastic flying machines and performed well on floats, too.

Jay had purchased a new Cessna 185. But in a few months he was crying on my shoulder, unhappy with the new plane. He had discovered that while the 185 hauled a bit more weight, and was a little faster than the 180, the 185 aircraft couldn't get off the ground as quickly on a short runway.

Jay wondered if he might be able to buy back the 180.

"Sorry, Jay, no deal," I told him.

Chapter 9

Jimmy Burns

Big Lake, 1958

Jimmy Burns was a likable guy who navigated through life with a rain cloud over his head, just like the character in the Li'l Abner cartoon.

The handle on Jimmy's ax broke. His dog knocked a bowl of soup out of his hands. You get the idea. I happened to be working nearby one day when Jimmy, an outboard motor clutched in his arms, missed a step moving from the dock to his boat. He went down feet first with a splash. Weirdly, when he hit the lake bottom, still upright and weighted down with the outboard still in hand, he simply turned around and *walked* underwater up onto the beach. This reminded me of the 1950s movie, *Creature from the Black Lagoon*. I was transfixed as he calmly set down the motor and wiped his eyes.

Just another day in the life of Jimmy Burns.

Jimmy and his wife Agnes, who owned a beauty salon in Anchorage, were regulars at the Yacht Club. Jimmy loved my mother's spaghetti and meatballs.

From all appearances, Jimmy was an ordinary fellow—average height, one-hundred-sixty-five pounds, brown hair, and forty-something—who blended into a crowd. I thought he might have a career as a bank robber. But as ordinary as Jimmy seemed to be, he had an uncanny ability to get into extraordinary situations.

Jimmy took bad luck in stride, smiling as if he had no worries. He would invite you to walk through his garden and you would follow, even though you knew an alligator was going to leap out from behind the green beans and bite you in the ass at any moment, or you might simply fall off a cliff behind the monster cabbage.

So, one day when Jimmy opened a conversation with, "Chuck, guess what I just bought?" I put aside what I was doing, grabbed some coffee, and sat down. This should be interesting.

"What did you get into now?" I asked with eager anticipation.

"I bought a Republic Seabee amphibian airplane," he announced proudly. "It's over in King Salmon."

"Hot damn!" I shouted. "That's terrific. When will you pick it up?"

"In a week or two. I've got a mechanic going over to check it out. He's going to fly it back to Anchorage."

I was happy for Jimmy, but doubtful. I was glad to hear someone else was going to fly the Seabee before he got his hands on it.

Two weeks later, Jimmy walked into our lodge with a beaten dog look on his face. I could not bring myself to ask what happened. But I sensed it was bad. We sat down together, neither of us speaking for at least five minutes.

"The plane vanished on the trip back," he said. "No sign of it along the route."

I walked around the table, putting my hand on his back. What can you do at moment like this? I looked outside. Sure enough, a black cloud was passing overhead.

The wreckage of Jimmy's Seabee was found several years later, the mechanic still at the controls, in the bottom of a dry riverbed in Southwest Alaska. Most likely the pilot had encountered weather problems and was attempting to make a 180-degree turn at a low altitude, possibly to make an emergency landing, when he crashed into the boulder-filled stream bottom. I thought about this a lot because I had owned two Seabees. Neither had worked out for me, either.

Another year passed with no new disaster, so Jimmy took a leap of faith and bought a second plane, a Stinson 108 Station Wagon on floats. The first time he taxied up to the lodge, Jimmy stepped off onto the sandy beach beaming as only someone with a new airplane can.

"What you think of this one?" he asked.

"Not bad, James, not bad," I said.

"Get in and take it for a spin," he said.

I fired it up, and Jimmy gave it a push and a pat. I taxied around the lake a bit, checking things out. I'm no Catholic, but just before opening the throttle, I thought about having a rosary around my neck. But there were no surprises. In those days, we were accustomed to flying underpowered planes. The Stinson was powered by a 165-horsepower Franklin engine, a reliable engine, but the new Cessna 180 had 230 horses.

I eased the Stinson up onto the step, accelerated to flying speed, lifted the right float out of the water, and flew.

"Nice plane," I said to Jimmy.

For the next few months, my friend came and went in his Stinson. He seemed to have honed his skills. He was flying like a pro.

* * *

HUNTING SEASON and bad weather in Alaska both arrive late in the fall. Stories of adventure and close calls are plentiful at Peggy's.

Pilots liked to sell their wives on the idea that the annual hunt is vitally important because it puts meat on the table for the coming winter. But the reality, wives discovered, was this: the cost of a pound of wild game could run into thousands of dollars when you add in eggs and bacon, propellers, pancake mix, mosquito repellent, fuel, ammunition, perhaps a new set of floats, beans, coffee, and the cost of a helicopter to retrieve the broken plane. Of course, these expenses did not include medical bills and legal fees for the occasional divorce.

I was not surprised when I heard Jimmy Burns was missing in the Alaska Range, where he had gone to hunt Dall sheep with a companion. An underpowered Stinson on floats would not have been the first choice of most pilots to chase sheep. But then, Jimmy was Jimmy.

The first bad news came from Jimmy himself. His plane was equipped with an old Lear transceiver radio. Somehow he raised an FAA station some one-hundred-fifty miles away. Jimmy reported wrecking his plane on top of a glacier, but was uncertain of his exact position. He and his passenger were okay, but they were short on food.

I was cussing Jimmy out. So were others who knew and loved him. *Really, Jimmy, no food?* Jimmy's flying friends and CAP volunteer pilots fanned out into the mountains. The weather was typical for hunting season: sky obscured, one mile visibility, fog, and rain.

Jimmy made a second call on his radio, describing his location in more detail, and the search area was shrunk to a manageable size. However, the weather was worsening.

Eight days later, the crew of a DC-3 spotted two men on a gravel bar at the foot of a glacier. They circled back for a closer look. By then the two men were jumping up and down, waving their arms frantically. Word was relayed to the Rescue Coordination Center that the missing aviators apparently had been found. A helicopter was dispatched carrying a rescue party and two buckets of Kentucky Fried Chicken. Jimmy told me the story later.

"We were fed up sitting up on the glacier freezing our butts off. Our sleeping bags were wet. We had no way to build a fire. But at least we had plenty to eat."

"What are you saying?" I asked. "We heard you had no food."

"Well, I figured if I said we had nothing to eat, it might hurry up the search." Jimmy said. I shook my head in disgust.

"The weather improved before we were found," Jimmy said, "we had been sitting on a couple of big rocks, sunning ourselves as if we were on Waikiki Beach. No shirt or shoes. We were warm and dry. The helicopter landed about a hundred feet away. The crew was having difficulty walking in bulky winter gear, including insulated rubber boots. They handed each of us a huge bucket of Kentucky Fried Chicken.

"While we sat half-naked on the rocks, gnawing on chicken, the crew stomped around trying to stay warm. They thought we had gone crazy."

"I wouldn't have thought that, Jimmy," I said. "I always expect weird things to happen when I'm around you. That's normal. Give me the whole story. What caused you to crash?"

Jimmy said he had flown through Rainy Pass, about a hundred-twenty-five miles northwest of Anchorage. "I was thinking I could put the plane down somewhere on the north side of the range and cut the climbing distance to the sheep. But the weather kept getting worse, even though we had an 8,000-foot ceiling. Then the engine began icing up, so I opened the carb heat a bit."

This was a mistake. "Jimmy, you should have used full heat until it cleared up. You were making the situation worse. What happened then?"

"I was flying over the top of a glacier trying to climb," Jimmy said. "The engine wasn't doing too good, so I opened the carb heat all the way out. Then the engine quit. There was no time to react. I was only about fifty feet above the ice, so I turned downhill.

"This is when things started happening fast. I hit the ice with the right float, bounced, and then hit with the left float, which came off. We spun around, busted the other float, and then sheared off a wing. We were bouncing from crevasse to crevasse until we came to a huge one. The fuselage, with one wing dangling off, slammed to a stop in the bottom of the crevasse. I knew then we were in trouble."

"*You knew you were in trouble?*" I exclaimed, "I can't believe you survived. I know those Stinson airframes are tough, but how did you get out of the crevasse?"

"We crawled out of the cabin and surveyed our situation," he told me. "The sides of the crevasse were nearly straight up, about twelve feet high. The wing was torn off, so we pushed it around into position, cut the fabric next to the ribs, and turned it into sort of a step-ladder. I figured the first thing I would do is see if we could get the radio out of the wreck and try to make it work."

But apparently lunch came first. "We crawled down our ladder, pried open the baggage door and pulled out a big pot of spaghetti my wife had made for us," Jimmy said.

"You crazy bastard, I'm still angry at you for saying you didn't have anything to eat," I said.

"With our bellies full," Jimmy said, "I got out my little canvas bag of tools and went to work. First we got the battery out so we wouldn't fry any of the radio gear.

"Good thinking."

"We had in mind to build a little shelter up on top of the ice by using the cowling and the doors," Jimmy continued. "We pulled these parts up with a piece of rope I took off the float. Then came some floorboards to set the radio gear on once the shelter was in place."

I was excited by the thought of Jimmy pulling this off.

"We took out the transmitter, receiver, load coil, power supply, and all the wiring," he said. "I also pulled out the trailing long-wire antenna and reel. We stuck a float paddle into a crack on the surface of the ice and jammed the float pump into a crack fifty feet away."

"I don't believe it. Jimmy, you're a genius. How did you tune the antenna?"

"It took us a long time to hook it up, but once I had it right, I strung the wire between the wooden pump handle and the paddle, then wrapped a wire around the long-wire antenna and slid it from side to side while keying the microphone. I kept my eye on the tuning meter so when it peaked I hollered into the mike. I got hold of the FAA in Kenai and told them where I thought I was."

"Then what?" I asked.

"I waited a few hours before I called again, wanting to report a more precise location. But it was raining and foggy. I couldn't see far enough in any direction to get my bearings, plus the battery was low."

"What happened next?" I asked. I was expecting all sorts of mishaps, knowing Jimmy's history, but never in my wildest dreams was I prepared for what I heard.

"We walked around in a wide circle for five or six days," Jimmy said, "always scanning in the distance and listening for an engine. Nothing came.

"Then, one day I noticed a dark spot below my feet. Something was frozen in the ice. It looked like an animal. The rain was washing the surface of the ice so it became easier to see. *Oh, my God!*"

Jimmy said it was an enormous animal, and its head was close to the surface. "I got my hunting knife and began to chip away at the ice," he said. "The first thing I exposed was part of a tooth. I kept at it until the entire tooth was exposed. Chuck, you won't believe it—the tooth was five inches long, curved, and sharp.

"It was a sabertooth tiger, intact and completely frozen," Jimmy said, his voice cracking, "and I know exactly where it is."

When he got back to Anchorage, Jimmy contacted the U.S. Fish and Wildlife Service and later heard from a museum that had gotten wind of the find. An officer from Elmendorf Air Force Base wanted to know about possible landing sites for a big helicopter near the crash site.

"I guess this is going to be a big deal," Jimmy said. "Those guys should be able to find it because I took precise compass bearings off the rock walls on both sides of the glacier. I've got the location nailed down tight and marked out for them."

From barbershops to bars, Anchorage soon was buzzing with news of Jimmy's discovery and questions and speculation about how the big tiger came to be frozen in a glacier. Meanwhile, Jimmy was thriving on the notoriety as he made the rounds of Rotary Club meetings and dinner parties.

I was impatient and wanted to get in on an expedition. "Jimmy, what's going on? Time is flying. I'm worried about your tiger. Glaciers are slow, but they do move, you know."

"Chuck," he said, "this thing is getting bigger by the day. It's out of my hands. I can't even find out who's in charge."

* * *

"THEY CAN'T FIND IT," Jimmy muttered, his face buried in his hands.

He had shuffled into our lodge, flopping into an old overstuffed chair. He was horribly disappointed. I thought about pouring him a shot of whiskey, but he wasn't a drinker. I didn't know what to say.

Three months had passed. A considerable amount of planning for the tiger's extraction from the glacier was all for naught. The glacier had continued its slow movement downhill, grinding everything in its path. Nature had reclaimed its prize.

A few years passed without additional grief finding Jimmy Burns. He seemed happy enough, though he had no airplane. How can a pilot be truly happy without a plane? When word circulated around Big Lake that Jimmy was missing, my heart skipped a beat. Surely my friend soon would appear at the lodge and order a plate of Mom's spaghetti and meatballs, a wide smile in place, eager to tell another story of survival. Alas, it was not to be.

Jimmy and his wife had a summer cabin at Big Lake with a twenty-five-foot dock for their boat. Searchers found him drowned in eight feet of water off the end of the dock. A scuba tank was strapped to his back, and no one knew why the air supply failed.

Chapter 10
Doctor "Patch 'Em Up"

It was about ten o'clock at night; somebody was pounding on the door of the lodge. I got up, crawled downstairs and flipped on a light. I opened the door. There was a guy standing there covered with snow. "What's up?" I asked. "We have a buddy here who is hurt. How about taking a look at him, maybe you can help him out."

"Put him on the couch over there and show me what we've got to deal with," I said. They stretched him out on our old sagging couch in the dining room of the lodge. His leg was bleeding near the knee; it looked like it might be a big one so I said, "Let's cut his pants leg off so I can get at it." Sure enough, there was a cut across the front of his knee from side to side and deep too! This looked serious and required immediate action.

In those days there was no 911. It was 50 miles into Palmer where the nearest hospital was available. Since I was the only person in the local area who had been to college, everyone assumed that I must have had some medical training. So they were forever bringing me their children to bandage up and even give them their post delivery shots. I picked fishhooks out of eyeballs, sewed up stab wounds and put splints on a variety of broken bones. I knew most of the doctors in Anchorage in those days. They kept me supplied with all the bandages, simple drugs and ointments they thought I would need. So it wasn't much of a surprise to be awakened in the night and handed a wounded snowmachine driver.

Andy Anderson was a friend of mine so I looked him in the eye and said, "this thing is going to take some major stitching - probably five or six to pull it together. It'll hurt like hell so what do you want to drink? - whiskey, rum, Scotch and vodka is all I have, so what's your pleasure?" Andy, who was obviously in serious pain, chuckled and said, "something expensive."

I asked my dad to uncork a new bottle of Chivas Regal Scotch whiskey which was very expensive. The bottle came in a gold velvet and tasseled sack. Everyone in attendance said in unison, "ME TOO!"

I went into the storage room and began to gather up the equipment I would need. In those days we had a dog team, so I had a harness sewing kit consisting of triangular shaped needles, some curved needles, waxed thread and a sailor's palm. The palm was used to push the needle through the canvas, or in this case through the thick skin on the front of the knee. Ann gathered up some soap and towels and an old bed sheet to tear up as needed. We washed all the surgical tools in 190 proof booze called Everclear.

I was now set, ready to ply my new profession of "saw bones." I thought about all the old swashbuckling movies I'd seen where they put some poor hapless bedraggled sailor up on a table and whittled his leg off while he bit down on a stick. I thought about giving Andy a stick to bite down on but by this time he was well lubricated with Chivas Regal.

With Ann as my surgical nurse assisting me I put in seven stitches. This required using the sailor's palm to push the needle through the thick skin till it popped out. Then I grabbed the needle with a pair of pliers and pulled it the rest of the way through. I used the curved needles and regular. thread on the two ends of the wound. It looked pretty good for an amateur job, so I bandaged it up and patted Andy on the head while congratulating him on being such a good patient.

As they were leaving I told Andy to see a doctor in Anchorage in about a week so he could have the stitches removed. "Which doctor should I go to?" asked Andy. "Aw go see Dr. Renn on L St. She's a friend of mine and I know she'd be happy to take care of you." About a week later Andy walked into Dr. Renn's office, past a sign that read OB/GYN. There were eight or nine women sitting in the waiting room. Andy approached the nurse's desk, who looked up somewhat surprised and asked? "May I help you?" "Yeah," said Andy. "Is your wife a patient?" inquired the nurse. "No, I need to see the doctor." The nurse tilted her chair back to get a better look at Andy. "Why do you need to see the doctor?" "Well, it's like this," Andy said. "I need to have my stitches taken out." The ladies in the waiting room who had been listening in on this erupted into laughter. Andy was the only one in the room who didn't get it.

Dr. Renn looked over Andy's hem stitched knee and asked, "Who did this job?" "Chuck Sassara," he answered. "Not bad, not bad at all," said Dr. Renn. "I'll have to add Chuck's name to my letterhead, along with OB/GYN." This story cost me a lot guffaws and several Margaritas the next time Dr. Renn showed up at our lodge.

One summer night there was a big crowd swirling around our lodge. People were launching boats off trailers, lugging boxes of food and other assorted gear down to the dock. Kids were dashing all over the place. The bar was hopping, with my dad and brother Dick who was home on leave from the Army mixing drinks and popping lids off bottles as fast as they could. My mother, Katie and Ann were manufacturing steak sandwiches in wholesale lots. The whole scene was near bedlam, when someone comes running in a panic looking for me. "What's going on?" I asked. The man was excited but managed to spill out the story.

He said, "We were loading our boat which was tied to the dock when my friend fell down. He had two cases of beer in his arms when he stepped off the dock. His foot slipped and one leg went inside the boat and other outside. He fell straight down straddling the edge of the boat, landing with full force on his testicles. "Holy moley," I exclaimed. "Is he all right?" I asked. "No he's not all right. We have him lying in the bottom of the boat but we can't tell what's going on with him. His eyes are rolled up in his head so all you see is white and he is rigid and shaking all over. He might be dying! Come quick."

I had been helping the ladies in the kitchen, so I took off my apron and jogged out to the dock. I jumped down into the boat and took a quick look at the victim of this accident. I turned to the two fellows who were accompanying him and said, "Get him out of the boat and into that Super Cub over there right now." The Super Cub was on floats pulled backwards up onto the beach. It belonged to one of our regular customers.

They had a hell of a time putting him into the back seat of the Piper Cub. He was so rigid with pain they couldn't fold his body so he would be in a sitting position. I yelled at them to quit worrying about details and get out of the way. As I fired up the engine I again yelled at them, "Call Providence Hospital in Anchorage and tell them to meet me at the North end of Lake Hood."

It was pitch dark when I landed on Lake Hood. There were no lights in those days so you had to set up for the landing, holding a very steady rate of descent a bit nose high and using some power. The real trick to a successful landing was not to try to guess where the surface was. Just stay in that attitude until the floats kissed the water, then chop the power.

The poor soul was still rigid with pain when the medics pulled him out of the plane. I circled around 360° looking for any other traffic, then poured the power to the Super Cub. All takeoffs on floats are done by feel but of course being in a Super Cub it was in the air so quickly you really couldn't screw it up. I flew back to Big Lake where I had to dodge boats that were dashing this way and that. I landed and taxied up to our beach, turned the plane around and pulled backwards up the beach and secured it. A week or so later I ran into the owner of the airplane and told him I had borrowed it. It was not an unusual thing in those days. Nobody took the keys out of their airplanes, trucks, automobiles or locked their cabins scattered around Alaska. Somebody might need to use it.

The poor guy was in the hospital for a week or so. We never did find out how it all turned out. Everyone wanted to know but no one wanted to ask.

Whitey's guide service up in Eureka offered hunts for moose, caribou and sheep. A hunter from New York wrote to Whitey and arranged to be guided on a sheep hunt. Whitey later told me that he had misgivings about crawling up a mountain with his client. The poor guy was way overweight and not in any shape to climb a mountain, nevertheless Whitey put him in a Super Cub on floats and flew him in to the base camp. Up the mountain they went, Whitey leading the way. Somehow or other the New York sportsman managed to top a ridge at 8,000 feet. He looked around at the landscape and laid down and died.

Whitey really didn't have any options. He retreated back down the mountain, climbed into his floatplane and headed over to Lake Louise where he could enlist a couple of other guides to help him pack the man down the mountain.

There were no helicopters or rescue services available in 1950s Alaska. He managed to get three other guides talked into helping him retrieve the dead client. They took two float planes and an Army surplus litter which they tied onto the float struts. Someone suggested that they should alert the territorial police down in Palmer that they would be bringing in a body.

It was just another day in the life of Bush pilots and guides. But things became complicated when they tied the man on top of the litter and started down the mountain with four people carrying it down steep rocky terrain, slipping, sliding and stumbling like a three legged centipede, the guy getting away from them, sliding off the litter, whacking his head on stones and getting various scrapes and scratches. They didn't give up, just stopped and put him back on. Finally after about an hour of this, the four of them sat down, lit up some cigarettes and discussed another plan.

They had decided that getting him down that mountain in one piece was just too big a job. So the discussion was which way to cut him up so they could pack him out on their Trapper Dan packboards. They had convinced themselves that it wouldn't make any difference since he could be sewed back together and with a new suit and tie he would look okay for the funeral.

The litter was causing more problems than it solved so they decided to ditch it and just do the best they could dragging him down the mountain. They wrapped a spare jacket around his head, put extra gloves on him and began the descent with the dead hunter traveling feet first. When they got him down and next to the floatplane they had another problem. They couldn't place him inside the plane. In Alaska it is common practice to carry external loads on airplanes. Two by fours and sheets of plywood are carried on top of the floats. So they pulled him over to the float and the four of them rolled him up on top. They lashed him down tight because they sure as hell didn't want to lose him after all this effort.

When they arrived at Lake Louise there was a committee waiting for them including a territorial policeman, driving a station wagon with a metal body container. There was a coroner and some man from the CAA. The four pilot/guides who thought they had done a yeoman's job were being hammered by officialdom. The coroner was checking the body over for signs of foul play, like for instance, bullet holes. The gentleman from the CAA wanted to see their pilot's licenses and the territorial trooper was asking for their guide licenses. The three volunteers turned on their buddy and said, "Next time forget me, I don't need this!"

It turned out that Whitey, who advertised with a huge sign alongside the highway which read in big letters WHITEY'S FLYING SERVICE, didn't have a pilot's license. Only one of the four had a commercial pilot's license - that was Doc Pease. Ralph Marshall who lived on Big Lake near me only had a student pilot's license. He was famous for getting lost. Every time we turned around a big search would be started for Ralph. Somehow or another he was beloved by his clientele and his girlfriend who owned a bar in Anchorage. She bought him a new airplane each time he wrecked one. The CAA was pulling its hair out so they laid down the law: "You guys go get some licenses now!"

Chapter 11
Joe Redington

Big Lake, 1958

We had a team of sled dogs tied out behind the lodge. Some had been given to Ann and me by Charlie MacInnes, who had been hired by Pacific Northern the same day I went to work for the airline. Charlie's wife, Kit, was one of Alaska's leading sled dog mushers. The dogs were Siberian Huskies, a working dog bred by the Chukchi people of northeastern Siberia that were imported to Alaska during the gold rush.

Other neighbors on Big Lake began bringing dogs over to contribute to our team. We started with two dogs, then we had six, then ten. It was a lot of work feeding and exercising them. Our team feasted on food from our restaurant—salmon, sourdough pancakes, T-bone steaks, and spaghetti, mixed with dry dog food and enough water to make a good stew. Our dogs were known for their red muzzles from our Italian sauce.

One year I decided to enter a team in the annual Anchorage Fur Rendezvous Sled Dog Race in February. Ann, who loved animals, thought this was a great idea. Figuring we would need more dogs, I asked around to find out where I might borrow a few. Someone suggested I contact Joe Redington over on Flat Horn Lake. I knew Joe, having met him out on the trail from time to time.

How would I find Joe's place? Just fly northwest from Big Lake, I was told. You'll know you've found Joe's place when you look down and see a bunch of dogs in a huge dog lot near the lake. You can't miss it.

One morning I donned my mukluks, gloves, and down jacket, then stopped by the pot-belly stove in the lodge where I had stashed a bucket of aviation oil to keep it from freezing. This allowed me to get the Aeronca started and into the air fairly quickly. Another trick we learned was to drive the plane up onto some spruce boughs or short chunks of wood in cold weather so the galvanized metal bottoms on the wooden skis would not freeze to the ice. This was before a fellow in Anchorage invented the Landes fiberglass skis.

Soon I was flying across the Matanuska-Susitna Valley looking for Redington's old army tent where he, his wife Violet, and sons Joe Junior and Raymie lived. The first thing I saw, though, was a whole raft of circles in the snow created by dogs chained to their little shelters—usually simple boxes with a flat roof. The roof was always flat because the dogs liked to sleep on top, even in subzero temperatures, though a little dry straw spread out inside the boxes would entice them to come in for a nap occasionally.

I landed on the frozen lake and taxied up to the tent surrounded by seventy-five circles. The dogs greeted me with a chorus of howls and yowls.

Vi Redington invited me in for coffee and a snack consisting of a rolled-up sourdough pancake. Every day she would climb out of bed and start making a mountain of pancakes for the family.

At seven-thirty a.m., it was sourdoughs with bacon or an egg.

At ten, it was sourdoughs rolled up with a layer of sugar.

At twelve noon, it was sourdoughs with a bowl of moose stew.

At three-thirty, it was sourdough pancakes smeared with peanut butter.

At six, it was sourdough pancakes and salmon.

At ten p.m., the kids went to bed with sourdough pancakes sprinkled with sugar and cinnamon.

In between pancakes, Joe I looked over his lot of eager volunteers. The dogs were jumping and lunging in anticipation. I swear they sensed there was a plane ride for those who made my team. Within a few minutes Joe had pointed out three siblings.

The little girl in the group was a bundle of energy, her batteries always fully charged. She was a blur. I swore I never saw her with more than two feet on the ground at any one time. Her name was Rocket.

Her brothers were Jet and Comet. One was mostly black and the other sort of gray. As soon as Joe introduced them I forgot which was which. For the next two years I called one Jet 'n' Comet and his brother Jet 'n' Comet. The dogs didn't seem to mind.

I thought I was ready for the Fur Rendezvous race until I began chasing my dog team around the course. The three-day, timed sprint race was twenty miles on Friday, twenty on Saturday, and twenty-five miles on Sunday. I got into a rhythm of pushing the sled, then running a while, then jumping on again, riding for fifty feet or so, then jumping off and pushing some more. It was discouraging to discover that the winners rode most of the way around.

I finished sixteenth out of thirty. Only the winner and the last-place musher were remembered. The last-place finisher was presented with a red lantern since it was often after dark by the time he crossed the finish line.

I learned a lot from Joe Redington, who was a serious musher. He dreamed about running a marathon sled dog race from Anchorage to Nome, a distance of more than a thousand miles.

Joe had teamed up with another buddy of mine, Ray Genet, a mountain climbing guide. One spring the two of them decided to climb Mount McKinley with a dog team. Ray showed Joe a route up the mountain and damned if these two men and Joe's dogs didn't make it to the top.

Joe died in 1999 at age eighty-two after fulfilling his dream. He became known as the "father of the Iditarod" by co-founding the world-famous Iditarod Trail Sled Dog Race with Dorothy Page of Wasilla.

Ray froze to death on top of Mount Everest in a futile attempt to save the life of a women climber who was stuck near the summit. He had been guiding a group of mountaineers when he came across the stricken woman, who had been left in a sleeping bag, alone, by her climbing partners

Ray decided to bring the woman down the mountain, though he knew the odds of both of them surviving were not good. Neither made it.

Chapter 12
Trouble on the Mountain

Anchorage, 1960

My brother Dick was discharged from the Army at Fort Richardson in Anchorage. He had survived his tour of duty by serving as a ski patrolman in the winter and a hunting and fishing guide in the summer at a rest-and-recreation center for officers and visiting Congressmen on Lake Louise, about two-hundred miles northeast of Anchorage. As far as we could tell, the Army had given Dick a two-year paid vacation.

Dick's next employment opportunity presented itself when he ran into John Fields, an old acquaintance from Panama. A communications expert, John had worked for Air America, a front for the Central Intelligence Agency, at various places around the world. When he stepped off a flight to Anchorage, the first thing to catch his eye was Mount Susitna, also known as the "Sleeping Lady," a hulking flat-top of a mountain across Cook Inlet about forty miles northwest of Anchorage. A light went on in his head.

John went to the Piper dealer at Merrill Field where he hired a pilot to fly him around the top of Susitna looking for a place to land. The pilot happened to be George Kitchen, who had a lot of experience in Super Cubs. After a few passes, George found a place to put down on top of the 4,396-foot mountain. The narrow landing spot was just two-hundred-feet long, and littered with rocks and shale, but he was confident he could get in and out.

George dropped John off on a sandbar alongside the Susitna River and flew back to the mountaintop for his first landing. As expected, it was a bit rough, George recalled later, but he got the Super Cub down with room to spare, then applied himself to the task of rolling big rocks off the crude airstrip at what became John Fields' new "trade and manufacturing site." This is what it was called by the federal Bureau of Land Management when John leased federal acreage on the summit.

John's move was brilliant. From this high perch atop the Sleeping Lady, radio signals could be heard and transmitted over thousands of square miles, north to Mount McKinley and south to Homer, including the entire Matanuska-Susitna Valley, Anchorage, and the Kenai Peninsula.

A few days later, John hired a helicopter pilot, Link Luckett, to haul some two-by-fours and plywood to the site. Then he had Link haul in a small generator, an antenna, guy wires, and somebody to keep an eye on things and make sure the generator was kept full of fuel. This hardy somebody was Dick. My brother sat up on the mountain for months. When he got bored, he would walk up and down the airstrip picking up rocks and pitching them over the side. George Kitchen was kept busy flying in five-gallon cans of diesel fuel and supplies. Years later, George told an Anchorage newspaper reporter that he made four-hundred- eighty landings on the summit of Mount Susitna.

After establishing a transmitter on the mountain, John opened a sales office in Anchorage where he signed up customers for a two-way radio service. He kept asking me if I wanted a radio in my car. I didn't think I needed one, but when John offered me a free radio, I couldn't refuse.

Meanwhile, John had purchased his own Super Cub on floats. He wanted me to take his employees hunting. Ten or twelve people were working for him by then and were eager to get into the air. This led to another offer from John that I could not refuse. For each hour I spent flying his employees around, I could use the Cub on my own for an hour. Life was good. Now I had the Aeronca Sedan for wheel flying and a shiny new floatplane tied down in front of our Lodge on Big Lake. Everybody was happy.

I was still selling commercials on KENI radio and TV and doing a little broadcasting. This was supposed to be a winter job but I was still at it in June. But this was okay. Business at the lodge was slow and we needed the cash.

The stations were owned by Al Bramstedt, a beloved and widely respected pioneer broadcaster who visited us often at the lodge. Bram had one of those distinctive radio voices that made people stop what they were doing and listen when he went on the air.

* * *

ON MAY 17, 1960, a beautiful spring day, I was driving around Anchorage making sales calls when I heard an urgent plea for help on my new two-way radio. I pulled over to listen. The caller was a climber near the summit of Mount McKinley trying to raise the National Park Service or military. John Day, a wealthy sportsman from Oregon, had been injured in a fall. Day was part of an expedition that included world-class climbers Jim Whittaker, the first American to reach the summit of Mount Everest, and his twin brother, Lou, both from Seattle.

I called Al Bramstedt and told him what I had heard. Bram didn't hesitate. He recognized a major story was breaking.

"Chuck," he said, "switch to a reporter's role. Find out what's going on. Can you fly up there? We will cover the costs."

I raced over to Merrill Field, topped off the tanks in the Aeronca, and embarked on an adventure I'll never forget.

In the spring, which is prime time for mountaineering in Alaska, the weather is typically settled and clear. I could see the 20,322-foot Mount McKinley one-hundred-sixty miles to the north. Never having flown higher than 8,000 feet, I stopped by Casey Fesperman's office to borrow an oxygen tank and mask. Casey was an aerial photographer who had spent years mapping Alaska. Soon I was approaching the massive mountain, overwhelmed by its majesty.

The farthest north of the world's great mountains, McKinley is so tall it creates its own weather. I saw no lenticular clouds forming on the downwind side of peak, nor any powdery snow spin-drifting off the ridges. Conditions could not have been better for taking a look. Unfortunately, neither my Aeronca with its 145-horsepower Continental engine, nor I as a relatively inexperienced pilot, had what it took to get up to 25,000 feet. We ran out of climbing ability at about 15,000. I turned around before getting into trouble, but I was able to assess the situation from the lower altitude.

All hell broke loose the next day when the national news media picked up the story.

In 1960, there was no way to rescue climbers from McKinley except to put on your gear, grab a bunch of rope, and head up the mountain hoping you could pull off a rescue without injury or loss of life.

A rescue headquarters of sorts was established at the Summit airstrip alongside the Alaska Railroad tracks between Talkeetna and Cantwell. The only access was by rail or air. An Army rescue group had set up camp complete with a soup kitchen for volunteers who were beginning to arrive in anticipation of sending up a rescue expedition.

I flew in to Summit to see what was going on and report back to KENI. What a chaotic mess. A DC-6 had brought in about fifty would-be rescuers. Some were wearing Army surplus gear they brought in Seattle. Those who had a little climbing experience in the Lower 48 were packing an assortment of camping and hunting gear. Red wool jackets and mittens were everywhere. The Army personnel were mostly wearing Korean War-era gear. The place looked like a set for a TV comedy.

A big twin-rotor helicopter flew in. I was unable to keep a straight face when out stepped the perfect leading man for this comedy—a rotund, cigar-chomping U.S. Army captain named Norton. He surveyed the new encampment. I shook his hand and asked, "Do you think your helicopter can get up there?"

He removed the cigar from his mouth to take full measure of the stupid civilian who had asked such an impertinent question.

"No sweat," he barked.

From that point on I called the captain "No-Sweat Norton." Soon I'd seen enough and flew back to Anchorage.

On McKinley, meanwhile, the situation had gone from bad to worse. I learned that a second group of climbers was in trouble. An Alaska climber, Helga Bading of Anchorage, was suffering from altitude sickness at 16,000 feet. It was feared she would die if not brought down soon. I called Link Luckett, who had flown the building supplies to Mount Susitna in a Hiller 12E helicopter.

"Link, I don't know anything about flying helicopters but the conditions on top look favorable—clear with no wind. Maybe you could give it a try."

"I've been thinking about giving it a shot," Link said. "I called the Hiller Aircraft Company in California and asked them how high I might be able to hover. Taking into account temperature and wind, they figure I might get up to 10,000 feet. That's a bunch of altitude for a 200-horsepower piston engine, you know."

"You can say that again," I said. "I was just hanging on the prop when I was up there yesterday."

"Tell you what," Link said. "I'm going to check with my insurance company to find out if they will cover a landing on the mountain."

"You do that," I told him. "I'll be heading over to Talkeetna. I've got an oxygen system you can use."

Back at Merrill Field, I ran into Bernie Shaw, owner of a new Piper PA-23 Aztec Twin, a six-passenger plane powered by a pair of 250-horse-power Lycoming engines. Over coffee at Peggy's we discussed the situation. Bernie was full of questions. Mostly he was looking for an opportunity to put his new airplane to work. He volunteered that the Aztec could get up to 23,000 feet okay, but he couldn't stay there long without oxygen.

About then, a *Life* magazine photographer contacted Al Bramstedt. Bram referred him to me. I told the photographer about Bernie and his new Aztec.

Things were beginning to come together, and none too soon. Helga Bading had been lashed up in a sleeping bag. Understanding the difficulty of lowering the stricken woman 1,800 feet down a rock face to a potential landing site for evacuation, a volunteer pilot tied an Army-surplus ahkio sled onto the outside of a Cessna 180, flew the sled up to 16,000 feet, and cut it loose near where Helga's party waited.

Then, Paul Crews and another climber, Chuck Metzger, secured Helga in the fiberglass sled and began an emergency descent, lowering the woman with them. This was tricky business. The two men were putting their lives on the line. But they managed to lower Helga down to a glacier at about 14,200 feet. This was the only place anyone thought an airplane might be able to land.

At the same time, the Whittaker party was still hunkered down at 18,000 feet. The Alaska Air National Guard tried to help. Its mission was to drop food and extra tents from the rear of a C-123. For one reason or other, however, none of the supplies landed close enough for recovery by the climbers. Either the parachutes did not open or they overshot their targets. I watched this from five-hundred feet away flying in the Aztec with Bernie Shaw and the photographer. *Life* magazine featured many dramatic photos in its next edition.

We were circling the top of the mountain making left turns when a C-123 showed up in our windshield at the same altitude, making right turns. We were nose to nose and closing in. He dove hard left and shot down the mountain. Later, on the ground, I talked to the pilot. He had seen the high tail on the Aztec and thought we were a Fairchild F-27 closing in on him fast. We shook hands and laughed about it.

Back at the Summit airstrip, meanwhile, the military was babysitting seventy-five or so volunteers who were walking around in circles, drinking gallons of coffee, and standing in line for free food. "No-Sweat" Norton had loaded his crew into his big helicopter and headed out in hopes of rescuing *somebody*. He was last spotted sitting on glacier tailings at about 4,000 feet. Empty ration boxes and cans were scattered everywhere.

We landed in Talkeetna, sixty miles away, where Link Luckett was preparing to attempt the impossible. He had started the engine of his helicopter as we helped remove the battery and doors. Link needed to unload the extra weight to get extra altitude. He also figured an injured climber could be loaded more easily without the doors. A small bottle of oxygen and four masks were carried in an aluminum suitcase.

About then, Talkeetna Bush pilot Don Sheldon took off in his Super Cub. Link followed in his Hiller. Talk about rolling the dice. The Hiller engineers thought Link's mission was impossible.

"If you survive the landing up there at 17,500 feet, you won't be able to lift off again," they told him.

I kept worrying about Link as the Piper Aztec gained altitude on our way back to McKinley. Link had no insurance. He had confided to me that when he contacted Dan Cuddy, president of First National Bank of Anchorage, Dan told him the bank had no objection to the rescue flight provided the helicopter was insured. The insurance company said "no way," but Link decided to go anyway.

I looked down at the whirling blades on this little helicopter, knowing Link was squinting into bright sunlight reflected from the massive face. Only a handful of people have ever faced such a challenge.

Bernie and I decided to keep our distance from the summit for fear of getting in the way, or triggering an avalanche. The day before, a pilot and his passenger had approached in a Cessna 180 for a look at the climbers stuck at 18,000 feet. The Cessna stalled out, began to spin, and then crashed and burned near the top of the mountain. Dr. Rod Wilson, a physician from Anchorage who was climbing the mountain, made his way to the crash site and confirmed what everyone feared: the pair had burned to death. Rod was the busiest climber on the mountain, trying to keep Helga Bading alive and tending to climbers with various injuries in his and other parties. He was a genuine hero and plenty tough, too.

We learned from radio chatter that Link had landed safely, taken John Day aboard, and managed to get back into the air. Link told me later he had flown the helicopter right up to the steep McKinley slope at full power. When he gingerly touched his skids to the ice, the helicopter began sliding backward down the slope, but miraculously came to a stop a few yards short of the Whittaker brothers. The injured climber was quickly strapped into the passenger's seat.

Link never reduced power the whole time. When he pulled on the controls to lift off, the craft began sliding again. Suddenly he slid off a precipice. The helicopter was in free fall, but he regained control.

Link had set the record for the highest landing and takeoff ever attempted in any aircraft.

* * *

FARTHER DOWN the mountain at the 14,200-foot level, Paul Crews was watching over Helga Bading. All of a sudden, a Super Cub on skis appeared. Circling, it lined up for an uphill landing at the top of the glacier, pulling all the power he could from the 150-horsepower Lycoming engine. The pilot landed at a dangerously steep angle, a very unstable situation, but used lots of power to control the aircraft and managed to get it secured. The pilot was George Kitchen.

George knew Paul Crews well, and the two men embraced. Both had pushed their limits. And their ordeal wasn't over yet.

George prepared his plane for Helga Bading, who was trussed up like a cocoon. He had made room for her by removing the rear seats and rear control stick and putting down a pad in the baggage area. They were about to load her aboard when suddenly another Super Cub landed on the glacier, turned sideways on the slope, and slid to a stop. Out jumped Don Sheldon, hollering. At first George could not make out what the fuss was about, but soon it became clear Don wanted to evacuate Helga back to Talkeetna, where the news media awaited.

"Get her out of George's plane and put her in mine," Don said. "I want to take her down."

"I just shrugged my shoulders and said, 'okay, okay,'" George told me later.

While Don Sheldon and Paul Crews lugged Helga over to Don's airplane, an argument flared up between George and Don about best use of flaps at high altitude. They couldn't agree whether it was better to set half flaps on takeoff, or pull them when needed, like we did at sea level. George, who put a thousand hours a year in his Super Cub, was typically a quiet man. But on the glacier that day he became angry and disgusted. The hell with it. Sheldon could do it any way he wanted.

At this point, George loaded another climber, started his engine, poured on the power, and raced down the glacier in a cloud of snow, setting a record for taking off in a fixed-wing aircraft at high altitude—14,200 feet.

Chapter 13
A New Opportunity

Big Lake, 1960

Business was booming at Big Lake. We thought about expanding the lodge—going big, opening a fine new restaurant and a thirty-room hotel. Everybody liked the idea, even our bank. However, I was hesitant to commit to an ambitious, long-range plan as I wrestled with three issues.

First, we had two boys to raise. In 1960, they were ages one and four. The closest school was twenty miles away. Winter temperatures commonly dropped to thirty or forty below, and occasionally fell to minus fifty. The Big Lake kids stood out by the road in all weather, no matter how cold, waiting for the school bus. This worried me.

Second, I wanted nothing to do with the bar. Mostly I stayed away from it. The bar was tended by Dad with help from my brother when Dick was around, while Ann and my mother did the cooking. At any rate, I couldn't stand to be around a drunk. Slurred speech makes me recoil. Yet, because the nearest Alaska State Troopers detachment was in Palmer, thirty-five miles away, I had to step in when there was trouble. I had no patience for this, so there were a lot of fights. I tipped the scale at two-twenty-five. I knocked out some nasty drunks, but took no pleasure in doing so.

Third, Alaska had become a state. I was itching to join in the discussion about shaping the new state and to become involved politically. Big things were happening.

The upshot is we sold the lodge and moved into Anchorage, where I had continued to work ever since we settled at Big Lake. In the city we bought a sprawling log home on an acre and a quarter in a nice neighborhood called Rogers Park.

I looked around for a way to put food on the table. There it was—a large chunk of property on the west end of Merrill Field across the street from Peggy's Airport Café. I saw opportunity here.

I leased some ground near a hangar, put up a small office building, and then stuck steel poles into the ground to hold up all the flags and streamers I thought we needed to attract attention to my new endeavor—Airport Auto Sales. We sold cars, trucks, and airplanes from the same lot. Word got around I would buy, sell, and trade up, down, or sideways. I would even take a car or truck as trade-in for an airplane.

The new business was an immediate success. Pilots came out of the woodwork to check me out, kick a few tires, and find out what sort of deal they could make. Often I would walk across the street with a prospective buyer and make a handshake deal over pancakes and eggs. Trade in an old straight-eight Buick and taxi away in a four-place Stinson. GIs from Fort Richardson bought most of the used cars and trucks I took in trade.

My first week, a prospective buyer walked in and settled into a chair, announcing his desire to buy a four-place airplane. Larry Thompson was a fireplug of a man, about five-foot-three without his logging boots. He pulled out the thickest wallet I ever saw, bulging with odds and ends of paper scraps and frayed, old business cards. For the life of me I couldn't imagine sitting on that lump all day.

We walked out to look see what was on the lot.

A Cessna 170? Too much money, Larry said.

An Aeronca Chief? Too small.

A Stinson 108 Voyager recovered with new aluminum skin? "Yep, that'll do," Larry said. "How much?"

"I'll let you have it for $3,200," I told him.

"I'll be back by noon," Larry said. Right on time, he returned with eight checks of various denominations, apparently scrounged from friends.

Off he flew. I was relieved when the last check cleared. I was showing somebody a clean Piper PA-12 Super Cruiser ten days later when Larry showed up in my office with a frown on his face. He slumped into a chair and told me in a hoarse whisper that he had wrecked the Stinson.

"It wasn't the plane's fault," Larry said. This was no surprise. Most accidents involving small aircraft are caused by pilot error.

"I was landing on the main road down on the Kenai Peninsula, a two-lane gravel stretch with a big dip that drops about two-hundred feet. There was no traffic. I thought a landing would be easy—just touch down near the top, roll down the slope, and then stop on the uphill side."

I couldn't believe what I was hearing. "Larry, you crazy coot, you can't stop a plane running downhill."

"Yeah, I found that out. When I touched down in a three-point attitude, the plane quickly picked up speed. When I hit the brakes, the plane spun around and went off the road backward into a drainage ditch. That's when it flipped upside down."

"Wow!" I exclaimed. "That would have been good enough for a highlight reel. What's left of the Stinson?"

"Not much," he said. "The engine is okay, but every piece of that plane is bent."

"What's next?" I asked.

"I need another airplane," he said.

"How about that sweet Cessna 170A I showed you before?"

"I don't have enough money for the 170, right now," Larry said, "but I think I could scrape up $1,800 for the Aeronca Chief."

"Done deal, Larry. You gather up the money and I'll check over the plane and write up the paperwork."

The next day, Larry was back with a cluster of third-party checks and a fistful of cash.

* * *

YEARS LATER, Larry sank a Cessna 180 on floats in the Innoko River in western Alaska. He left four German hunters stranded for almost two weeks while he bummed a ride with a boater down the Innoko to the Yukon River, where he caught a flight back to Anchorage. When he arrived in my office, a look of desperation on his face, I pulled an aircraft order form from my desk. Larry told me his sunken airplane was still tied to a tree, nose down in a cut bank.

By then I was calling him "One-Way Thompson"—fly out and walk back.

Larry took more than a week to raise the money for a new airplane to retrieve his clients, for whom he had left a handwritten note. The stranded Germans had been eating nothing but moose and beans. I heard on the grapevine they were upset and angry at first, but after showers and a hot meal back in Anchorage, they convinced themselves that the ordeal actually had been an adventure. The men even paid full price for the charter and gave Larry a tip.

Larry took a scuba tank, a compressor, and a pile of large inner-tubes back to the Innoko. He shoved the tubes inside the floats and pumped air into them. Up came the Cessna, water and mud pouring out everywhere. After sloshing the plane with river water, he replaced the spark plugs and changed the oil and fuel. Last, he installed new magnetos.

When he twisted the starter key, wonder of wonders, the engine fired up. Larry flew the 180 back to Lake Hood, where he tied it down and turned a couple of garden hoses on it and in it. It took forever to wash out the mud.

Poor old One-Way Thompson crashed the last Cessna I sold him while hauling dog food for the Iditarod race. He hit a mountain head-on in bad weather. We mourned his passing.

* * *

MOM AND DAD, in their mid-sixties and full of energy, were not content to slip into retirement after we sold the lodge. One evening they came over to announce they had bought airline tickets to Paris.

"We thought we would wander around Europe for a few months, especially Italy and maybe Spain," Mom said.

"Ann, let's go," I said. It sounded like fun. The timing was good because the boys, by then three and five, were not yet in school. I had a partner of sorts who would look after the business while I was away.

In Paris, the first thing we did was have lunch in the Eiffel Tower. Then we became tourists for a few days. For serious travel we needed a car, and everyone said Germany was the place to get one.

Two friends from California, Ed and Beverly Isensen, were working for a Swiss news bureau in Bern. I left the family there and caught a train to Munich. In no time, I was headed back to Switzerland in a four-door Mercedes-Benz, a shiny black sedan with rounded fenders made out of thick steel and long hood.

We rolled away across Switzerland, Germany, Austria, France, then down the length of Italy. We had no travel plans, no itinerary, just a few maps and guidebooks and a new guest house every night. From Calabria, at the toe of the Italian mainland, we ferried over to Sicily to visit the birthplace of my mother's parents.

Campo Felice was an ancient village halfway between Messina and Palermo. Like the rest of Sicily, it was accustomed to invasions—the

Greeks, Phoenicians, Romans, Normans, American soldiers in World War II, and now the Sassara clan.

The entire village turned out for a celebration to greet us. The people were open and friendly, and there was so much food we could not begin to eat everything placed before us. I was surprised to find my Sicilian cousins, aunts, and uncles were mostly blue-eyed and red-headed. Seems the Normans had been a major influence here.

Next, we loaded the Mercedes onto a freighter headed for Tunisia. From there we sped across North Africa, dodging camels, hordes of sheep, and bullets.

Mom had spoken her Pittsburgh Italian in Italy. In Spain, my dad used the Spanish he learned wandering the length and breadth of South America from age thirteen to twenty-four. I took on the job of communicating as best I could with the Tunisians, Algerians, and Moroccans. I picked up a few words of the local dialect, smiled a lot, and didn't stay too long in any one place.

The U.S. embassy in Tunisia was uncooperative when I asked for help getting a visa to drive cross Algeria. They said there was a war on—the Algerian revolution. But the Algerians had been fighting for years to win independence from France, so I was not worried. I found an Algeria-interest section in the French embassy. Ours were the first visas they ever issued.

Rolling through Algeria, we saw a lot of barbed wire, a few shot-up cars, and schoolhouses that appeared to be abandoned. Ever the optimist, my mother noted it was five o'clock in the afternoon and suggested that everyone must have gone home.

As we neared the Algeria-Moroccan border, a car sped up alongside our Mercedes on a deserted two-lane road. Inside were four Algerians waving pistols and motioning us to stop. No way. I stomped the throttle and the chase was on. Our German car carried a heavy load—four adults, two kids, and our baggage. The crazy gunmen were driving a Citroen sedan.

I pushed the car to one-hundred-fifteen miles an hour. As it turned dark, Ann, who was following our progress on a map, warned me about a hard turn coming up. This was a break for us. As we disappeared around the turn, I slammed on the brakes and the Citroen flew by, a near miss. I lost our pursuers in a nearby town where the Algerian army had a checkpoint backed up with tanks and soldiers.

It was pitch dark at the border. When I shut off the engine, we heard rifle and machine-gun fire all around us. The Moroccan Army was spread out along the border, its mission to keep out Algerian war refugees. The Moroccans were shooting at anything trying to cross the invisible line in the sand.

Inside the little border station, twenty or so Arabs were sitting on the floor, wrapped in hooded cloaks. A charcoal brazier was providing warmth. A radio was playing weird music. No one spoke. It took thirty minutes to get our passports stamped and entry documents signed.

Outside, the sky was awash in stars. I lifted my eyes in awe, wondering why these people were trying to kill one another. It seemed so pointless in the overall scheme of things.

"You okay?" I asked Ann as we crossed the border.

"I could use a couple of Rolaids," she said with a smile.

After six months of travel, I handed Dad the keys to the Mercedes, which was still bright and shiny and running like a fine watch. He and Mom had decided to linger a while longer in the house we had rented in Malaga, Spain. Ann, the kids, and I flew to Los Angeles by way of Copenhagen with a fuel stop in Greenland. When we walked through Customs in LA after fourteen hours in the air, Charlie complimented the U.S. Customs agent who was looking through our luggage.

"You sure speak good English," little Charlie told him.

Back in Anchorage, we put Charlie in school. Ann joined a women's club. Richard and his best friend, Brad Stern, armed themselves with wooden rifles, helmets, and water bottles. Their mission: defend our front yard from invaders.

I went back to trading airplanes.

* * *

ORGANIZING THE new state of Alaska was an enormous, complex task. A bright and dedicated group of legislators and new state employees were wading through the issues, resolving some, setting aside others that required money Alaska did not have. Maybe later. Construction of major projects such as office buildings, schools, airports, and ferry boats required voter approval of general obligation bonds.

Governor Bill Egan, our state's first elected governor, appointed me to an advisory council for the federal Small Business Administration (SBA).

This came about in part because of my reputation as a small businessman and in part because of connections through my brother Dick, a registered stock broker who was well known in the banking community. The appointment was timely because I was positioning myself to run for the legislature. But I knew winning a seat would be a mixed blessing. I'd moved into Anchorage to get involved in politics, but lacked the steady income I would need to support a family while spending six months a year in Juneau for the legislative sessions.

One day in October, 1962, a black sedan came roaring onto the lot next to Merrill Field. Out jumped Governor Egan. He ran into my office, out of breath, gasping, "Chuck, grab Ann and your boys right away; there may be a war starting."

"What do you mean? A *real* war?"

"Yeah, a real war," he insisted. "The Russians are putting long-range missiles in Cuba, aimed at the United States. Get your family out of here. Find a safe place to ride it out."

Having delivered his warning like a modern-day Paul Revere, Egan hustled back to his car. Why the governor decided to warn me remains a mystery.

I had a Cessna 170B parked next to my office, outfitted and ready to go. Ann and I discussed what we would need to survive for six months. We both knew exactly where to go—Polly Creek, one-hundred fifty miles down the west side of the Cook Inlet. It had a nice sandy beach, fresh water, and a thirty-foot tide that exposed clam beds and flatfish in the puddles. There was plenty of firewood and straight timber. Moose and bears were found in abundance. All we needed were nails, hand tools, pots, and pans. We would set up a campsite, then build a log cabin.

In eight years Ann's perspective had gone from Hollywood glamour to backwoods survival at Polly Creek. She was at ease carrying her own custom-made 30.06 high-powered rifle looking for a moose to plug, or a bear to chase off the porch. This woman was always ready for a new adventure.

The Russians backed down, nuclear war was averted, and we missed an opportunity to go camping.

Chapter 14
The Merry Little Band

Anchorage, 1962

I loved working and meeting people at the northwest corner of Merrill Field, where a community of flying services and aviation-related enterprises, including my own, were thriving just across the Glenn Highway from Peggy's. The café was a perfect gathering place for pilots. Peggy was a short, jolly, and ample woman who knew how to make pies and sourdough pancakes. This and barrels of coffee were all it took to keep our little air force operational.

Oren Johnson was a short, earnest, bespectacled fellow who never met anyone he didn't like. He reminded me of a school teacher standing in front of a black board, pointer in his hand. But his quiet demeanor clashed with the reality of his life. Oren was a fiercely competitive stock car driver who regularly traveled on the local race circuit.

Between races, Johnson flew a Piper PA-22 Tri-Pacer and managed the property he leased to members of our merry little band.

His tenants included Barton Air Service. Bill Barton spent a good deal of his time sprawled on a beat-up couch in his office, occasionally crawling off to fly a charter. Bill didn't look like your typical pilot, either. He had a belly that sagged as badly as the couch and eyeglasses with lenses as thick as Coke bottle bottoms.

Another flying entrepreneur, Bert Johnson, operated Johnson's Air, which specialized in chartering a couple of Bell 47 helicopters. Bert was a blur, always on the move. One day he told me something I've never forgotten.

"The economics of aviation is simple," Bert said. "If you can see your plane through your office window, you're going broke. It's only making money when it's in the *air*."

Slim Walston ran the propeller shop. Straightening props was big business in Alaska. Some mornings I saw a cluster of pilots leaning on their mangled props outside of the shop, shooting the bull while waiting for Slim to open.

Pilots Ruth Obuck and Warren "Ace" Dodson flew for Barton Air Service. Ruth later went on to work as an inspector for the FAA, married another inspector, and retired in Anchorage in 2008. Ace became a captain flying Boeing 737s for Wien. Ace's father, Jim Dodson, was a well-known 1930s Bush pilot. Ace died in a crash while flying charter for a camera crew filming the Iditarod race.

Fred Notti, who had been our next-door neighbor at Big Lake, kept the helicopters flying at Johnson's Air. Fred was a small, dark, and handsome man, half-Italian and half-Athabascan Indian. He was an aircraft mechanic courtesy of the U.S. Air Force. He had no pilot's license, but he flew anyway. It seemed the FAA spent an awful lot of time investigating Fred, hoping to catch him in the act. An inspector came out to the lodge occasionally to question me about Fred's flying habits.

"I don't know anything," I told them. "Which one is Notti?"

Life was never dull around Merrill Field. One day Ruth Obuck asked if she and a hunting partner, who were flying together in a Super Cub, could borrow my Piper Tri-Pacer. Ruth had shot a large bull moose on the other side of Cook Inlet and faced a daunting task of getting the meat out.

"We are on a sandbar in the middle of the Drift River and need two planes to fly it out," Ruth explained. "We had landed next to the river and tracked the moose. When I got off a shot, the big bull took off, wading into the river. The blasted thing managed to get up onto a smaller sandbar, then turned around to look at me and dropped dead."

"I left my partner there cutting it up. I'll bet that moose weighs more than 1,200 pounds. I'm going to borrow a 180, too, and with your plane, maybe we can get it home in just one more trip."

Soon they were off, Ruth in a Cessna 180 and Oren Johnson flying my Tri-Pacer. Two hours later, they returned with an enormous pile of moose meat on board the 180. When Oren stuck his head into my office, a sheepish look on his face, I knew something was wrong.

"Tell me your story," I said, "did you bend my airplane?"

"No, but I left it there sitting on a short sandbar. I landed okay, but when I got out and saw how little room I had to get it back into the air, I decided I'd better not try. Sorry."

"Don't sweat it," I said. "I'll ride back with Ruth and see if I can get it off the ground. I have a few tricks up my sleeve."

The Tri-Pacer is a tough little bird, way underestimated by pilots who've never flown one. Peninsula Airways in Bristol Bay operated Pacers and Tri-Pacers for years, hauling passengers and freight out of tiny fields, off of sandbars, and even flying off the end of cannery docks. Those guys knew how to fly. If you wanted to know about Tri-Pacers, you checked with George Tibbets or Oren Seybert over at Bristol Bay

Back to Ruth's dilemma: She and I flew back over the inlet and landed on the little sandbar, pulling up next to the carcass of her moose. She had not exaggerated. This was an enormous moose. We piled most of what was left of the meat into the 180 and the rest of it into my Tri-Pacer. I was plenty heavy but not at gross weight. I thought it looked okay for a takeoff and decided to go for it.

First I taxied back as far as possible until finally I had to turn around, my tail hanging out over the water. Then I poured on the power, grabbed the flap handle, and held back on the wheel. I kept one eye on the airspeed, the other on the rapidly shrinking sandbar ahead. When the airspeed began to quiver, it was time to haul it off. With full flaps, back all the way on the elevators, the tail of the little Tri-Pacer smacked the ground and jumped into the air. The angle of climb was extraordinary but we were airborne. Finally I eased the flaps a bit and slowly dropped the nose. We had gotten off with five or six feet to spare.

Back at Merrill Field, eager members of our merry little band had rounded up a big barbecue stove made from a fifty-gallon drum split lengthwise. Wood and charcoal were found. Everyone had the same thought: moose steaks. I never did find out how Ruth's moose suddenly became community property, but she didn't complain. There was plenty to go around. The moose meat was dragged out of the two planes over to a large table where a butchering crew went to work.

Meanwhile, soda pop, paper plates, napkins, bread, and condiments mysteriously showed up. The word was traveling like a wild fire down the row of shops and hangars fronting the taxiway. An enthusiastic crowd watched as a fire was lit and the first chunk of meat was pitched onto the grill. Nearly everyone was packing a knife on their hip. This bunch was ready to pounce.

Then airplanes taxiing into position for takeoff began pulling out of line onto our gravel apron where they joined the growing crowd. About then I did a double-take when I happened to look across the runway. Someone in the tower was flashing alternating green and red lights toward us.

I yelled at Ace Dodson, pointing to the control tower. Ace turned and sprinted into his office. In a minute he was back, grinning like hell.

"What do they want? I asked.

"They want a ride across the active runway so they can have a moose sandwich," Ace said.

"Take the Tri-Pacer and get them," I told him.

The tower was almost directly across the runway. A couple of the air-traffic control operators were waving at us. I waved back and pointed to the Tri-Pacer that Ace was firing up.

John Arsenault, the tower chief, climbed out of the plane when Ace got back. Everybody liked John.

"Follow me," I told him. "I'll put you first in line. We weren't going to forget you guys."

"We weren't too concerned about missing out when we saw the size of that pile of meat," John said.

The impromptu feast attracted people and airplanes. Tower operators were ferried back and forth across the runway. Everyone told their best stories. I didn't count the diners, but figured we consumed more than five-hundred pounds of moose meat. As the September sun began to set, I looked over at Ruth Obuck in the center of a group of pilots. She smiled back. It had been a great day. The big moose had fed a small army, or, should I say, *air force?*

* * *

A FELLOW PILOT who would become a good friend walked in the door one day after I got a phone call from a salesman at an Anchorage real-estate company. The young salesman, the ink barely dry on his new pilot's license, had convinced the broker that the firm needed a plane to show remote properties around the state.

"What do you have in a comfortable, four-place plane that is easy to see out of?" the salesman asked.

I happened to have on my lot a North American Navion from the same company that built the Mustang P-51, a fighter plane flown widely by the Allies in World War II.

"I'll bet I've got just what you're looking for," I said.

Much as he wanted to fly, the salesman doubted his company would be able to insure an aircraft flown by a rookie pilot with few hours. But he had an idea. "Our office manager's husband is an airline pilot who might fly it for us until I'm a bit more experienced," he said.

The salesman arrived at Merrill Field with the pilot in tow, a man whom I came to know as John "Earthquake" McBride. That's what people called him, though rarely to his face. A retired Eastern Air Lines pilot later told me he had flown co-pilot with John, whose frequent rough landings earned him the nickname. When I first laid eyes on John, he was wearing blue, food-spattered coveralls and house slippers. He was an enormous man. I took his massive hand in mine and shook it. This was the start of a friendship that lasted forty years.

The three of us walked out onto the parking apron where I showed them the 1947 Navion, a solid, roomy airplane with big windows and a retractable canopy that gave it the look of a fighter. The engine put out only 205 horsepower, but its airfoil and oversized flaps gave it good flight characteristics.

When Earthquake John and the salesman climbed up onto the wing and slid open the canopy, I knew I had a sale. The salesman was eager to get his hands on the controls of this exotic aircraft. John, a retired military air-transport pilot, was impressed with the Navion's rugged look. The two men hurried back to the brokerage office to tell the boss this was the right airplane. Within a day or two, the purchase was completed and John was hired as chief pilot.

John, who had thousands of hours in just about every piston-engine airliner you could think of, showed up for a checkout. He and I crawled up into the Navion. I am big, maybe two-hundred-thirty pounds. John was a six-footer who weighed at least three-hundred pounds. Yet neither of us had trouble fitting into the front seats.

I handled the radio because John, accustomed to flying airliners with a co-pilot, had forgotten the standard patter for communicating with the control tower.

"Merrill ground, this is Navion November 763," I said.

"Go ahead," the tower responded immediately. "What do you need?"

I asked for permission to do three touch-and-go landings. When we were cleared for takeoff, I gave John a thumbs up. Down the runway we thundered. At about a hundred-thirty-five miles an hour, I told him to go ahead and lift off because we already were at *cruising speed*. If we didn't get off the ground pretty quick, we were going to lose a wheel or blow out a tire.

When we finally rotated, I pulled up the gear and John made a wide, flat turn as if we were flying a Douglas DC-7 at full gross weight. We were already out of the pattern.

"Navion 763," the tower called. "I thought this was going to be a touch-and-go."

I grabbed the microphone. "We will be back in about ten minutes and will re-enter on a downwind for Runway Six, hopefully for a touch-and-go." I said. By this time we were three miles south of the airport. The tower said okay, and I turned to John.

"This time let's see if we can keep it a bit closer to the pattern. By the way, she'll fly off the ground at about sixty-five miles an hour, so no need to stay on the runway until you exceed cruising speed."

* * *

A FEW YEARS later I sold an airplane to John in Florida. He and his wife, who worked for a taxidermist in Miami, decided they needed a Piper Apache Twin I had parked in Lakeland. We decided I would fly the aircraft down to the New Tamiami Airport and we would complete the sale there.

This time, I checked John out mostly around the cockpit, explaining the fuel system and emergency gear-extension system. By then I had checked myself out in more than one-hundred-fifty different models of aircraft. The fuel system and landing gear were two things I always checked out carefully before opening up the throttles in a strange aircraft.

In South Florida, we shook hands and I left with a cashier's check in my pocket. John called later to tell me about his first flight in the Apache.

"We went over to Nassau, then down to George Town in the Cayman Islands and back," he said. "Everything went fine until I got back to Opa-Locka Airport in Miami."

"What happened?" I asked.

"It was dark by the time we got back to Miami, and I was cleared for a straight-in to Runway Two Seven," he said. "I dropped the landing gear and a bit of the flaps and was pulling up the nose for the landing, still about twenty feet in the air, when I pulled all four throttles back."

"What do you mean, you 'pulled all four throttles back,' it has only two engines," I shouted, unable to contain myself.

"I realized that when we fell out of the sky," John said. "She hit the runway like an overloaded footlocker. I'm used to flying four-engine planes. In the dark I'd grabbed four levers together. Only thing is, I had grabbed two prop controls along with the two throttles, so I feathered both props and the engines died. Down we fell, a vertical drop of almost twenty feet to the pavement.

"I must have sat on the runway for ten minutes trying to restart an engine. The tower kept asking me why I was sitting out there in the dark. I didn't answer—I was too embarrassed."

I went to sleep that night still thinking about John "Earthquake" McBride and his huge paws.

Chapter 15
Fred Notti

Anchorage, 1962

One winter day Bert Johnson experienced engine trouble flying one of his two Bell 47 helicopters over Knik Arm, forcing him to make an emergency landing on the ice next to Elmendorf Air Force Base.

Bert calmly climbed out of the chopper, walked across the ice to the shore, hiked around the bluff, and made his way into the Port of Anchorage terminal area, where he bummed a ride over to Merrill Field. There, he found his helicopter mechanic, Fred Notti, who had been our neighbor at Big Lake. The two men picked up some tools and drove back down to the port, intending to retrace Johnson's route and out onto the ice to the helicopter.

"Nuts!" Bert shouted as they rounded the point. The helicopter was floating away on a piece of ice broken off from the shore ice due to the fluctuations of Knik Arm's notorious high tides.

Bert and Fred raced back to Merrill Field, where they pitched tools and rope into Bert's second helicopter. They cranked it up and pulled out straight over Peggy's. Bert was sweating bullets, afraid he would lose the expensive chopper that was at that moment floating around in circles in the Knik Arm. They found the downed chopper and Bert hovered over it while Fred lowered his tools down onto the ice and then slid down the rope himself to see if he could diagnose and repair the malfunction.

"I hovered around while Fred worked on the chopper," Bert said. "I saw him climb in and suddenly the rotor was starting to turn. He was firing it up. Then, to my amazement, he picked it off the ice. I yelled at him on the radio to put it back down, but by then he was out over open water. He was flying sideways, then backward, and spinning around—it was wild! I was sure he would kill himself."

The stick on a helicopter requires tiny movements with your fingers and I imagined Fred applying a death grip on the controls, yanking them all over the place.

"By this time," Bert said, "I was thinking about what I would say to Fred's widow when suddenly Fred got the Bell under control. The next thing I know he is headed for the beach with its two-hundred-foot bluff."

Bert knew if the blades struck the bluff there would be nothing left but a mangled pile of wreckage. "Damn if he didn't manage to set it down on the beach," Bert said. "I couldn't believe our luck—Fred for having survived and me for having the helicopter in one piece. I looked back toward the city docks to see if there was any traffic, figuring I could put down over there and give Fred a ride back to Merrill Field."

"When I turned back, I couldn't believe my eyes. The crazy SOB was winding up the rotor blades again. By this time I was speechless, which didn't matter because every attempt to communicate with Fred had failed.

"Off the beach he came, a bit wobbly as he headed for the docks. Now I was worried about the loading cranes. But Freddie climbed right over them heading for Merrill Field. There he made a nice tight turn to avoid crossing the taxiway and set the helicopter down on the apron in front of the office."

When he got back on the ground, Bert said, "I didn't know whether to kill Fred or kiss him."

This was Fred's first and last helicopter flight, so far as we know. But his exploits were discussed widely by pilots, insurance agents, and of course FAA inspectors. Complaints had been filed and reports written, but no one ever caught Fred flying without a license.

* * *

FRED BECAME active in Native politics after Congress settled the Alaska Native land claims. Once he joined a group of Alaska Native leaders, bankers, and financial advisors gathered in Washington, D.C. to discuss how to invest the Alaska Native corporations' settlement money. The only woman at the gathering was Gail Talmadge, wife of Herman Talmadge Junior, whose father was a U.S. senator from Georgia.

When Gail and Fred looked at one another, it was love at first sight. They literally ran off together and later married after divorcing their spouses. Gail, an interior decorator, was hired to work on the new Sheraton Hotel in Anchorage. The lobby featured an incredible jade staircase curving up to the grand ballroom.

The couple spent time in Georgia, too. One day Fred was driving along on Interstate 75 with the top down on his convertible when two rednecks pulled alongside, apparently suspicious of this man with a dark complexion tootling down the highway in an expensive automobile. The redneck in the passenger aimed a small automatic pistol at Fred's head.

This did not faze Fred. He reached down for his weapon of choice, a Smith & Wesson .44 with an eight-inch barrel, which he pointed right between the eyes of the redneck just four feet away in the next lane. The other driver made a violent turn to get away and lost control of the vehicle, which landed in a ditch.

Fred saw no need to stop.

Chapter 16
Good Friday

Anchorage, 1964

On a Friday afternoon in March, 1964, the floor of my office at Merrill Field suddenly began moving. Startled, I jumped up to grab a doorframe. Airplanes and cars were bouncing up and down. Plate glass windows at Peggy's were exploding into the street. I soon learned schools and apartment buildings were falling apart all over town and an entire neighborhood was sliding into Turnagain Arm.

At first, I thought the Soviet Union was attacking. The shaking continued for more than five minutes in some places. This turned out to be the Good Friday Earthquake, the second most powerful earthquake ever recorded, measuring a Richter 9.2. I jumped into the car and headed for home, weaving around debris and cracks in the road. Our sturdy log house was still standing. Ann and the boys were fine. We were lucky. The earthquake claimed a hundred thirty-nine lives. Property damage was estimated in the hundreds of millions of dollars.

Our only loss was a gallon jar of vanilla extract, which tumbled off a shelf and shattered. Vanilla splashed everywhere. For years the house had a sweet, somewhat familiar smell. "What's the scent?" visitors would ask.

After the earthquake, phone calls poured in from all over the world. Everyone wanted to know if we had survived. One of the first calls was from a worried local banker.

"Could you fly me down to Girdwood?" he asked. "My daughter is there visiting friends. The road is destroyed and I can't get to her. I've got to know if she's alive."

"I'll meet you by the tower at Merrill Field in twenty minutes," I said.

After learning that his daughter was okay, the same banker hired me to fly him to Kodiak, Seward, and Valdez for a look at the widespread devastation.

Anchorage International Airport was shut down. The earthquake had destroyed the tower and left huge cracks and heaves in the runways. Light planes were pressed into service.

Another banker chartered a flight to Valdez. There, I waited around until nearly dark for the return trip. Finally my passenger showed up, drunk. As we flew over Columbia Glacier, this dangerous idiot turned the ignition key and pulled it out, shutting down the engine. Then he opened the passenger window and was about to throw out the key, but I snatched it back and restarted the engine. By then we had dropped to about a hundred feet above the icy crevasses. Angrily I changed course for the Palmer airport. There, I yanked the banker out of the plane and left this idiot passed out on the runway.

Governor Egan soon called a special legislative session to address urgent matters related to the disaster. Here we were a state just five years old, still organizing the new government, and suddenly we faced a difficult and expensive recovery effort.

Unwilling to remain on the sidelines, I filed to run for a seat in the Alaska House of Representatives in 1964. I figured getting elected during this difficult time was akin to buying a ticket for passage on the *Titanic*, but what a challenge it was.

I had no clue how to campaign for office. One of my first moves was to write a position paper proposing construction of an Anchorage campus of the University of Alaska, which at the time had only one campus in Fairbanks. This proposal brought an immediate barrage of criticism. First, the idea was attacked in an editorial by Bob Atwood, publisher of the *Anchorage Times*, who just happened to be an Alaska Methodist University regent. I was criticized, too, by Dr. William Wood, president of the University of Alaska in Fairbanks, who was defending his turf. For similar reasons I was kicked in the face, so to speak, by Eugene Short, head of Anchorage Community College.

This was a bewildering lesson in down and dirty politics. But I fought back and continued to push the concept of a four-year state university in Anchorage, even adding to it a proposal for an Alaska student loan program.

Ed Isensen, a high school friend from California, and I made a sixty-second, black and white television commercial using an old sixteen-millimeter Bell & Howell camera. We shot it at sixteen frames a second, then played it at twenty-four, giving it the feel of an old silent film, then added some Scott Joplin piano music. In the commercial, my next-door neighbor, Gerri Ivy, played a damsel in distress. She runs through the woods, jumps into a boat, rows a while, and then climbs onto a bicycle and peddles down a dirt road to a bucket of paint. Stopping, she dips a brush into the paint, and marks a big X on a sixteen foot wide sign reading "Vote for Chuck Sassara."

The commercial created a sensation, receiving so much attention that viewers who had not seen it were calling the two TV stations wanting to know when it would be aired again.

Despite the rough start, the campaign was a wonderful experience. Shaking hands and making speeches was right up my alley, thanks to two semesters of public speaking in college. I loved the contact with the other candidates, too. In those days, Anchorage had fourteen at-large seats in the state House of Representatives and seven in the state Senate. U.S. Sens. Bob Bartlett and Ernest Gruening briefed us on issues of the day.

Gruening was especially helpful. At five a.m. one day, the phone rang, and the former territorial governor drop-kicked me into action.

"Why are you still in bed?" Gruening demanded to know. "Let's get out there and start campaigning."

I was elected.

I was sworn in with a marvelous cross-section of Alaskans in January, 1965. The new legislature included commercial fishermen, lawyers, Bush pilots, doctors, teachers, heavy-equipment operators, a liquor salesman, hunting guides, and an Eskimo whaling captain.

Taking it all in, I loved being a witness to history in the making as a Harvard-trained attorney debated a gold miner about how to build a state, and a truck driver explained to an orthodontist the need for state-funded kindergarten. This was an exhilarating experience, democracy in real time. There was talk about creating a model state government by learning from mistakes made by the Lower 48 states.

I liked and admired these people, whose bonds of friendship have endured for a lifetime. My colleagues in the Alaska Legislature were dedicated to the task of organizing Alaska's new self-reliance and making a better life for Alaskans. When somebody later asked me about this piece of history, I described the legislature as a boat with sixty sailors pulling together. Some had short oars and others long ones, but we all went in the same direction.

About half of the legislators were pilots, and many of us had flown to Juneau in our own airplanes. In the Senate, future Gov. Jay Hammond had been a Marine pilot in World War II, future U.S. Sen. Ted Stevens flew military transports during the war, and future Lt. Gov. Lowell Thomas Jr. served in the Army Air Corps. Willie Foster and Ray Christiansen were Bush pilots with many years of experience. Some legislators were hunting guides who flew their clients out to the Bush. Bill Poland from Kodiak flew his own airplane and operated a fishing boat as well. Often we would fly around the state to explain issues and legislative proposals.

Senator Gruening and his wife, Dorothy, invited Ann and me to a dinner party at his home in Juneau one night. The senator was beaming with pride. He had gone hunting, prepared the entire dinner, set the table, and poured the wine. Ann, who loved all critters, wouldn't allow me to shoot ducks or geese—or anything else, for that matter. While the rest of the guests were making lip-smacking sounds of approval, a frowning Ann was examining the lead shot she plucked out of a duck.

My favorite story about Senator Gruening described his response to a campaign tactic by Mike Gravel during a tight Democratic primary race for the U.S. Senate in 1968. Mike kept repeating, "Senator Gruening's age is not an issue," thereby calling attention to the fact that the good senator was eighty-four.

Senator Gruening promptly flew to Barrow, stripped off his clothes on the edge of the Arctic Ocean, and waded out into the frigid, ice-laden water for a leisurely swim—in front of news cameras, of course. I admired the man for being one of only two U.S. senators who in 1964 voted against the Gulf of Tonkin Resolution, which led to the Vietnam War.

Most evenings during the legislative sessions, Governor Egan would hold court in his office until the wee hours. I often went as a member of the House Finance Committee. The far-reaching conversations invariably turned to the crucial question of what federal land the new state should select as part of the 104 million acres promised by Congress in the Alaska Statehood Act.

The land grant was intended to give Alaska a source of revenue to be derived from mineral extraction and other development as the state struggled to become self-sufficient. Late one night everyone was bleary-eyed and sick of stale coffee when the governor introduced Phil Holdsworth, his commissioner of natural resources.

"Tell them what your geologists think we have up on the North Slope," Governor Egan said.

Holdsworth told us a story that was to change Alaska forever. "Gentlemen, there is oil up there. Lots and lots of it," he said.

Over the objection of some legislators, the governor selected the North Slope—a vast, flat, roadless plain between the Brooks Range and the Arctic Ocean. Critics believed there was nothing there except ice, snow, a few Arctic foxes, and some caribou. The more business-oriented legislators wanted to acquire land around Anchorage, Fairbanks, and Juneau, believing it could be sold off to developers.

In 1969, the state offered parcels of land near Prudhoe Bay in a competitive auction for the right to look for oil. The lease auction brought in $900 million. By comparison, the state budget for that year had been around $200 million.

Only a decade after becoming a state, Alaska had achieved financial self-sufficiency and faced a bright future.

Mom, Dick, age 6, and Chuck, age 9, are boarding a Sikorski flying boat operated by Pan American Airways. They walked out a gangway that led to the top of the plane then down a stairway to the cabin. We took off from Biscayne Bay, Florida and landed in the ocean in front of Havana, Cuba where we switched to a steam ship headed for our new home in the Panama Canal Zone October 14, 1940.

Our 1954 VW Microbus and Pappy's pickup in front of a "motel" on the Alcan Highway in 1955.

I liked this sign on the road at Lake Hood in Anchorage so much I stayed around for 60 years.

The new Anchorage International Airport is where I went to work two hours after we arrived in Anchorage. I was a passenger agent for Pacific Northern Airlines.

Charlie gets his first taste of moose - Christmas dinner 1956.

Here is our first house on Big Lake, Alaska. We cut down 160 trees, drug them out of the woods and peeled them using a small double bitted ax. We built a two bedroom house which we heated with wood burned in a homemade stove.

My mother bought this handsome four place Aeronca Sedan. It had no radios or instruments, nevertheless I put 600 hours on it flying in the Arctic and out the Aleutian Chain.

Charlie, age 9, and Richard, age 6, headed out into the woods packing 30.06 rifles. They came back grinning. "Hey dad, we got one."

Chuck draped on a Luscombe 8A, just like his first plane that he bought in 1951. Chuck told his instructor, "I'll never learn to fly this thing. It has too many instruments to watch: airspeed, altimeter and a compass!!"

Chuck and his neighbor Al Hibbard took two teams of dogs on a two week mushing trip near Paxson, Alaska in 1956. It was 35 degrees below zero and they lived off the land.

This is Rocket, the little sister of twin brothers Jet n comet and Jet n comet. Joe Reddington lent them to Chuck to fill out his team. Joe is the father of the Iditarod race.

Hal Clester and Johnnie Parks work on our home at Big Lake in 1957 with Charlie and Ann.

Dinner at the Yacht Club - a Piper J3 on Edo 1320 floats and a Super Cub on Edo 2000 floats sitting on the beach. I used both of them to guide in.

By 1956 I had gathered up enough money to buy a 1946 Taylorcraft model 12D on Edo 1320 floats. It could beat anything but a Super Cub into the air. We used to put three of us in it, along with our guns and camping gear. We would get it up on one float, run in circles and fly away.

Ann and Richard by the Stinson 108-2 that I sold to "one way" Thompson who wrecked it and walked back. I sold him three planes; he walked back from each one.

Richard wanted a closer look at the camels alongside the road in North Africa. We left Algeria in a hail of bullets.

This is my Cessna 206 on Chakachamna Lake were I parked it in a "white out." I knew where I was, but it took six days for the rest of the world to figure it out.

My Fairchild F24 W 46 was a real "arm chair" pilot's plane. It was powered with a Warner Super Scarab 165 hp. radial engine, flew great and was comfy to sit in hour after hour.

This was our second Globe Swift which I had painted in a desert camouflage and British Spitfire markings. The first time I landed at Miami International Airport, the ground controller asked, "Where are you headed with that?" "Sinai Desert," I answered, "Just stopped for fuel."

This is the PA18 95 that I put down on the first available spot during a massive thunderstorm. This caused a heated discussion between me and the head of the local FAA. Alaska won!!

A Piper Comanche 250 in a subdivision in Port Richey, Florida where I wanted to visit my father. I washed the mud off my landing gear and continued on to Northway, Alaska where it was 68 degrees below zero.

Ann loved to drive into Anchorage while wearing white gloves in her Rolls Royce. She often said "a girl ought to own a Rolls at least once in her lifetime." It was her staff car.

Chuck and Ted Stevens debating in the legislature - we got a kick out of getting everyone riled up over some issue. Then the two of us would leave and go get dinner.

Photo by Craig MacVeigh

Here is an Aeronca C-3 "bathtub" like the one the state of Alaska's first governor, Bill Egan, flew from Anchorage to Valdez in the winter of 1938. He made it in a blinding snowstorm, but it was very close.

This was my fifth Navion which I fixed up for my personal transportation.

Col. Wolfe Meister and Chuck on a Messerschmitt Me-262 twin jet fighter on display in the National Air and Space Museum in Washington DC. Chuck introduced Col. Meister to the head of the U.S Air Force. Col. Meister was a test pilot on the Me262 and later a wing commander in the Luftwaffe.

Ann and Sheba check out our Pilatus Porter. For lack of TV the Eskimos would line up along the village strip and watch me land in strong cross winds. I got even by threatening to open the bomb bay while I made a low pass, dumping their groceries all over the place.

Delivering diesel fuel - I finally started to carry four drums. The Cessna 206 would be overloaded by 600 lbs. Getting in the air was easy, but landings were like a lover's caress - softly, ever so softly.

Chuck Sassara, National Transportation Safety Board Chairman Webster Todd and his assistant had a meeting at the Bethel airport. Todd asked Sen. Ted Stevens "How can I find out what is going on with flying in Alaska?" Stevens answered, "Call Chuck Sassara up in Bethel; you will get straight answers from Chuck".

Ann loved DeHavilland Beavers. Ours had big tires and a 250 gallon belly tank. We delivered fuel to informal airstrips and on sea ice.

I'm in the front of this tail wheel version of the Varga Kachina. It had 0.32 skin, push rod controls and 180 horsepower. You couldn't catch it in a phone booth. George Varga paid me to demonstrate it to an Israeli general, the head of their flight training school. We had a ball and threw it all over the sky. Our only complaint was the lack of guns.

The Lockheed company had its engineers design a bush plane. Mine was built by Aermacchi in Italy and flown across the Atlantic. It had a monster door, swept back landing gear, and a turbo-charged engine. I bought this one in Seattle, flew it to Florida, then back up to Alaska. It never did find its niche.

Bob Vanderpool fueling my Cessna 185 using five gallon cans.
This is what a so-called FBO looks like in Alaska.

If you ever feel like you're in a Funk, this is what it looks like.

Brantly helicopter painted by Chuck in front of our new house in Girdwood. Our college-aged son Richard built the house for his parents.

Eskimo passengers in front of my old companion, an Aero Commander 560, which carried me down to the headwaters of the Amazon River - then up to Point Barrow, Alaska.

Here is our Curtis C46 resting in the weeds after both engines quit. Note the position of the props, one is feathered the other was turning at an idle. A month or two later we flew it away.

N611Z

Plane Crashes In Cow Pasture

The CallAir (Interstate Cadet) that I sold to "big deal" Bill Diehl. He liked the plane so much he started building these planes in Anchorage with 150 hp Lycoming engines. They are now known as Arctic Terns.

Chuck and one of his best pals, Oren Hudson, in front of Oren's Grumman Widgeon. Oren is the past president of the OX 5 club. The last time I asked, he admitted to having 36,600 flying hours.

Ann Sassara, former Hollywood débutante, who had stalked the wilds of Beverly Hills, takes those skills on a hike in Alaska at age 75.

Ann and I both agreed that this Cessna 195, our fourth, was the best. Turbocharged with a 350 hp. engine, it had a built-in oxygen system and would zoom right up to 20,000 ft. and push 200 m.p.h., then land on a dime.

Mom fueling guide Jack Lee's Super Cub. Note the cluster of antlers tied to the struts.

Pappy, Chuck and Aero Commander after flying from South America to Anchorage, Alaska.

The anchor in front of the Miami Yacht Club is the same one that I was pictured sitting on in 1936.

"Manakai" is one of the nine sailboats we have owned. It is a Bowman 36 built in England.

Ann at the controls of a big Bell helicopter in Point Barrow, Alaska.

I flew this Connie to Puerto Rico with the captain, John "Earth-quake" MacBride, snoring away. He didn't even want to watch my landing.

A Cessna 206 brings Christmas to the far, far north.

Seneca III that Melridge Aviation of Vancouver Washington, the Piper distributer, gave me and said, "Go sell airplanes."

Chuck and Ann at their 50th reunion at Los Angeles' University High School in 2000.

Chuck's cousin, James Irwin, who was on the moon driving the "Rover" when he found the oldest rock ever seen and brought it back to earth.

Chapter 17
Governor Egan

Juneau, 1966

Papers were scattered all over my desk on the floor of a quiet Alaska House of Representatives late on a Friday afternoon. Many legislators had flown home for the weekend. I stayed behind working on the budget.

A page summoned me for a phone call in the clerk's office.

"Who is this?" I asked.

"Bill Egan," he answered. "What are you doing still here? I understand your mother is in the hospital in Anchorage.

"That's right, governor, she's going to be operated on tomorrow."

"You should be up there with her," the governor said. "I'm flying up to Anchorage this evening, so stop whatever you're doing and I'll pick you up at your house at about seven o'clock."

How could I refuse? On our way to the airport, Governor Egan stopped at a grocery store, while I waited in the black, four-door Chrysler with license plate "Alaska 1." The governor returned laden with a bag full of supplies he said we would need for the flight.

At the Juneau airport, we pulled out onto the parking apron next to an Air National Guard C-123 cargo plane, which was powered by two big piston engines and built along the lines of a shoebox. It was fitted with a ramp that could be lowered to admit vehicles. On board, I laughed at the sight of a fifteen-foot camp trailer tied down to the deck. Governor Egan and I stepped into the trailer and settled in at a little dinette.

This was executive transport, Alaska-style. Once airborne, as soon as the C-123 leveled out, the governor broke out his paper bag, laying out bread, cheese, bologna, and an onion on the table, and began making sandwiches.

"Holy cow!" I exclaimed, impressed.

What a man of the people our governor was. Alaska's first elected governor and his brother, "Truck" Egan, had grown up in Valdez, a nice little port town on the upper end of Prince William Sound. Valdez and Cordova, a fishing town about forty-five miles away as the crow flies, were gateways to the gold and copper mining in the interior of Alaska. Truck Egan owned a liquor store and his brother owned a grocery store. Bill Egan had served in the territorial legislature, then ran for governor when Alaska became a state in 1959. A short, stocky man, Egan was a towering politician who had an uncanny ability to remember the names of nearly everyone he ever met and details of their lives, too.

Soon we were munching away and exchanging stories. The governor told me he was a pilot and had bought his first and last airplane in 1939.

"I didn't know that, what did you buy?" I asked.

"It was an Aeronca C-3," the governor said. "Got it from Jack Carr in Anchorage. I asked him to rig up an extra fuel tank, just in case I got lost or the weather turned nasty on the flight home."

"An Aeronca C-3, what a hoot," I exclaimed. "That thing looks like it was designed by Walt Disney for a movie cartoon."

Governor Egan laughed. "You are right," he said. "It was funny looking and, as I remember, the engine was only two cylinders and put out 36 horsepower. But it was a good flyer, at least that's what I was led to believe.

"When I got to Merrill Field, Jack walked me around my new joy and showed me the extra fuel tank he had installed. It held about two and a half gallons and was mounted to the floor between the pilot's legs."

The governor made me another sandwich, then continued his story. "A fuel pump from an old Ford V8 was mounted inside of the cockpit with lines leading out over the side of the airplane to the regular tank in the nose. It looked like it would pump fuel from the extra tank into the regular tank."

The governor said he stopped by the Weather Bureau in downtown Anchorage and asked for a forecast from Valdez. A Telex message was sent asking, "What's the weather like?"

A little while later an answer came back. "It's okay," the message read. Not exactly a ringing endorsement.

"I was set to go," the governor recalled. "All I had to do was to top off the fuel tanks, screw up my courage, and ask Jack Carr to pull on the prop. In this airplane, you didn't have to check much before takeoff. It only had a single ignition, and all you needed to know was whether it was running on both cylinders. Just pour on the power and go. At about twenty-five miles per hour you were off. You were only sitting a foot or so off the ground."

"Which way did you go," I asked, "around by Turnagain Arm, then over Whittier, or straight over the top of the mountains by way of the Knik glacier?"

"I planned to go by way of Turnagain," he said. "I didn't want to do much climbing. A guy could get a nosebleed up there at a thousand feet, you know."

By then, we both were howling with laughter at the image of Governor Egan heading out on a hundred-thirty-five-mile winter flight over forbidding and unforgiving terrain and water. This scenario would scare the pants off most pilots. There was no place to put down an airplane between Whittier and Valdez, a distance of more than ninety miles. We had to stop and wipe the tears out of our eyes and the mayonnaise off our faces.

"As soon as I got over Prince William Sound, I was in trouble," the governor said. "There were clouds everywhere and snow was falling. Things didn't look right. The usual landmarks and islands were not where they were supposed to be. I kept looking and looking for any point of land I recognized, so I could get my bearings. You know about that?"

I nodded my head in agreement. I'd been in this situation many times.

"By this time," Governor Egan continued, "I figured I better get to pumping fuel over to the main tank."

"It didn't take too long to find out that Jack Carr's design had two major flaws. One, I couldn't pump fast enough to keep up with the engine's need for fuel and, two, the pump fittings and hoses were leaking like mad. This scared the hell out of me. I thought I would blow up.

"I had been in the air over two hours and had no idea where I was," he said. "I was wishing I was home, stocking the shelves of my grocery store instead of flying blindly in bad weather. My legs were shaking uncontrollably. Suddenly, there was Valdez arm. I almost cried as I banked the little Aeronca and headed home."

"Governor," I said, "give me another one of those bologna and onion sandwiches." I had gotten so caught up in the story my palms were sweating and my mouth was dry.

"That's not all," he said, handing me another of his patented sandwiches and a Coke.

"There's more?" I laughed, and the governor continued his story.

"Well, you know Valdez," he said. "As I came in low over town, buzzing it so somebody would come to the airport and pick me up, I noticed about four feet of new snow had fallen."

"What did you do?" I asked.

"I couldn't go anywhere else with the little fuel I had left, so I just landed on the airstrip. The plane went a few feet and stopped abruptly. I unbuckled my belt and slid down out of the plane. The snow was above my waist. I could hardly move.

"Four or five guys wallowed their way over from town, picked up the plane and hoisted it over their heads and walked it off the strip," he said, suddenly standing up and flapping his arms. "Oh, how I loved that little bird."

Years later, after Bill Egan's death in 1984, I asked the governor's son Dennis what had become of the Aeronca. "My mom was not too happy about him flying, mostly because he was a bit of a daredevil and took too many risks," Dennis said, launching into a story.

As a kid, Dennis said, his dad had worked for Bob Reeve at Reeve Aleutian Airways, describing his father's job as "chief bombardier."

"They would load up the old Stinson Reliant with cargo to be delivered to the miners scattered throughout the mountains," Dennis said. "Often they couldn't land because the snow was too deep. So the drill was to fly the plane with the door off with young Bill Egan in the back—no tether, no parachute—digging through the boxes and dragging the right ones to the door and standing by until the pilot hollered, 'Shove it out.'"

Finally, Dennis answered my question. "With this kind of early training under his belt," he said, "my dad was considered a poor risk for insuring his Aeronca. So, when the hangar at the Valdez airport burned down, destroying the Aeronca, my mom said it was the best thing that ever happened to him.

* * *

ONE WINTER in Juneau, Governor Egan, legislators "Horrible" Harold Hansen, Nick Begich, Bill Moran, Bob Dittman, and I were having a few drinks in the Latchstring bar at the Baranof Hotel. We decided it was time for dinner and our unanimous choice was Mike's Place across the Gastineau Channel Bridge on Douglas Island.

Outside the hotel, we climbed into a cab, hooting and hollering, and set off for Mike's Place, where we ordered another round of drinks and dinner. Mike's Place was famous for its deep-fried crab legs and spaghetti. As we waited for our meals, the story-telling continued, rather loudly. We were having a grand time.

Then a state trooper walked in, took a look around, and made a bee-line for our table. By this time we were about the only customers left in the place.

"Who was driving the cab?" he asked.

Several of us answered at once: "The cab driver."

"No, he wasn't, he's outside in my car," the trooper said.

About then, the cab driver walked in. It seems he had parked his ve-hicle in front of the Baranof while he found a bathroom. The cab was gone by the time he returned.

"We've been looking for it the past thirty minutes," the trooper said. "Now, which one of you was driving it?"

We looked around at one another, trying to figure this out. I was pretty sure I hadn't been driving because I recalled being in the back seat. And I remembered Bob Dittman hanging out the passenger-side window in front, blowing on his saxophone. This narrowed the field of suspects.

Under state law, no legislator could be arrested while the legislature was in session. And the trooper didn't have the heart—or guts—to accuse Governor Bill Egan of auto theft.

"Let's get out of here," he told the cab driver.

Chapter 18
The Nuns

Saint Mary's, Alaska 1969

One of the many pleasures of flying in Bush Alaska was the constant variety of people and places.

Some days I would fly a bureaucrat clutching the ever-present brief-case trying to look important as he toured the government's remote outposts. The next day, my passengers might be a pair of elderly Eskimo ladies returning to their village from a shopping spree in the "big city" of Bethel, population 3,500.

Bethel was the Chicago of the vast Yukon-Kuskokwim Delta, a hub of transportation and commerce serving an area the combined size of Washington, Oregon, and Idaho. Not a single mile of road connected any of the fifty or so villages on the delta.

Travel was by light plane, dog team, or snow machine in winter, and plane or boat in summer. When I was majority leader of the state House of Representatives, having become familiar with the region working in Bethel for Christiansen Air, I felt an obligation to help address the problems of this thinly spread population when I could, even though I was elected from Anchorage.

The only steady sources of cash income for most people were winter trapping and summer fishing in small outboard-powered skiffs for salmon making spawning runs up the river. The Kuskokwim River fishery wasn't big enough to attract fish buyers from Seattle. Yet, in 1968, those U.S. fishing interests complained to Governor Walter Hickel when an Eskimo woman, Lucy Crow, helped local fishermen organize a cooperative and then negotiated a contract with a Japanese company to bring a processing ship up the river to Bethel.

The Japanese began buying fresh fish from individual fishermen for thirty cents a pound. The Seattle buyers were paying ten cents a pound elsewhere. Lucy Crow's co-op was on a roll.

One day I got a frantic call in Anchorage from Lucy. "The governor is trying to shut me down," she said. "He has sent state officials into Bethel, and they are walking up and down the river bank threatening to arrest anyone who sells fish to the Japanese ship."

"Hell's bells, he can't do that," I said. "Let me check this out. I'll find out what your legal options are."

My first call was to the U.S. State Department in Washington, D.C. I asked to speak with Henry Kissinger, then secretary of state under President Nixon. Of course, I didn't speak with Kissinger but apparently I did get through to someone high enough on the food chain to prompt an inquiry from the State Department to Governor Hickel.

Then I headed for Bethel. At the time I was flying a big, old Beechcraft B50 Twin Bonanza, a tough seven-passenger plane that was a little slow but performed well on rough airfields and strips. When I landed in Bethel, Lucy and her husband Bruce were waiting with a 1949 two-ton Chevrolet flatbed truck that sported a sign, "Bethel Fishermen's Co-op."

Our first stop was the river, where I went to work like an old-time preacher. The first person I ran into was a sergeant from the Alaska State Troopers.

"Do not tell these people it is against the law for them to sell their fish—it is *not* illegal," I told him. "If you persist, I will swear out a federal complaint against you."

I preached this sermon a few more times. Our bluff worked. Soon there was a mass exodus of state officials from the scene, cursing their bosses for having sent them to Bethel. The state had backed off.

The fishermen went back to hand-pulling their gill nets full of salmon and planning how they would spend the cash that soon would be burning a hole in their pockets.

* * *

BACK IN JUNEAU, during a legislative session, I was surprised to receive a letter inviting me to give the graduation address at the Catholic mission school run by the Jesuits in Saint Mary's on the lower Yukon River. I readily agreed, looking forward to being reunited with Antoinette Johnson, the Mother Superior.

Mother Superior was a handsome woman in her early fifties, not someone you would expect to be running a mission school in the Alaska Bush.

In years past, every time I flew mail to the lower Yukon River villages, I would make a point of walking up to the school at Saint Mary's to see the Jesuit priest and Mother Superior. I always asked, as I was leaving, did they need anything from Anchorage?

"Well, yes," she said one day. "I need to get to Providence Hospital for a checkup.

"Heck, yes," I said. "I've got a couple more stops upriver. I'll be back around five o'clock to pick you up, and then we will head over to Anchorage."

Mother Superior and I talked for four hours as we flew over the tundra and wound our way up and over the massive mountains and back down to Cook Inlet. She regaled me with colorful stories about living in the Bush, traveling by canoe, picking berries, eating wild game, and last but not least teaching reading, writing, and arithmetic to generations of Eskimo children.

A few days later, after the doctors turned her loose, I picked her up for the flight home. But as we pulled out of the hospital parking lot I had another idea.

"Hey, do you want to go fishing?" I asked.

She grinned. "Sure, I'd love to go fishing. I'm not needed back in Saint Mary's for another day or two."

"You will have a ball," I promised. "We're heading over to Whittier. I've got a big, comfortable boat and a skipper there waiting for you."

We drove the forty miles south from Anchorage, drove the car up onto an Alaska Railroad flatcar, and went through the railroad tunnel to Whittier. Waiting for us there was "Big Brownie," my skipper, who stood six-foot-five and weighed around three-hundred pounds. "Brownie" had mirth as well as girth, a rough-edged fellow with a heart of gold. I knew these two would get on well.

As I watched the two of them head for the dock, I had to smile, thinking of everything they had in common: a deep love of Alaska, an abiding faith in mankind, and an eagerness to reach down and help others less fortunate than themselves.

A few months later, I found myself stuck in Saint Mary's due to bad weather. There was nothing to do but wait it out. Learning that I was stuck, Mother Superior sent one of the kids down to the river to find me and send me back to the school.

"What do you mean you're going to sleep in the plane? You come on over to our house; I will find you a bed," Mother Superior said.

"Sleeping in a bed in the nuns' home would be nice, but will the nuns agree to this arrangement?" I asked.

"Sure they will. I'm the boss, you know," she said, laughing. "They will get a kick out of it. Having a man around the house and all."

As I settled in for the night in a real bed, saved from bending like a pretzel in the cockpit of a Cessna 185, I thought to myself, *someday I'm going to have to write a book.*

I was up early in the morning, trying to scrounge a cup of coffee in the dining room when Mother Superior came in. Over breakfast, she asked me an unexpected question.

"Chuck, considering your last name, I was wondering if you speak Italian?"

"A little," I said. "Why?"

"We have a nun here who was born in Italy," she said. "After we eat, let's go into the kitchen and say hello to her."

Sister Thecla Battiston was petite, about four-foot-six, an apron tied tightly around her tiny waist She was holding a pot, and her hands said she had known plenty of toil in her life.

I approached her from behind, leaned down, and whispered, "*Buon giorno, come è la vita qui?*" ("Good day, how is life here?")

Sister Thecla quickly turned around, raising her hands to her face, and began to cry. She hadn't heard her mother tongue in more than sixty years. She tried to explain that she couldn't remember how to speak Italian. I told her I had exhausted my vocabulary as well.

What followed was a three-way conversation among Sister Thecla, Mother Superior, and me, each of us trying to speak at the same time in broken English, broken Italian, and broken Eskimo. What a time we had.

As I made ready to leave, I told Sister Thecla that I would be back in Saint Mary's in about a week, and did she need anything from Anchorage? For the life of me, I couldn't think of what an eighty-year-old nun would need, but I always asked.

To my surprise she spoke up at once. "There is one thing I've always wanted," she said. "I would like to be a United States citizen."

This astonishing request left me breathless, and I had to sit down. Then I took this tiny woman's hands in mine and held on, turning to Mother Superior.

"Can you tell me what happened here?" I asked.

The story tumbled out. Sister Thecla had been a young nun born in Italy who traveled by ship to New York. After being processed through Ellis Island, she managed to find a train headed for Seattle, and then onto a steamship bound for Nome. From there, she was carried in an open longboat across Norton Sound and up the Yukon River to the mission.

"She didn't speak a word of English," Mother Superior said. "I don't know how she ever made it."

Still clinging to Sister Thecla's frail hand, my mind began to race as I digested this story. The course of action was clear. Sister Thecla would be granted her wish. Whatever mountain needed to be moved would be moved. I hugged both ladies and promised to return as soon as possible.

As I trudged down to the river, I began to formulate a plan. The first order of business would be to contact a federal judge in Anchorage. This turned out to be Ray Plummer, whom I knew from our Big Lake days.

"Your honor, this is Chuck Sassara, I need a big favor," I said.

"What's going on?" he asked.

"There is a Catholic nun over in the Saint Mary's mission who wants to become a U.S. citizen." I said. "She's been there for about fifty years. Could you go over there and set up court so we can help her out? She is old and frail and won't fly, so this may be the only way."

"That's amazing," the judge said. "I'll call my clerk and see if he will come with me. It would be an honor to take her oath of allegiance."

An hour later Ray called back. "My clerk is excited. He says he is ready to roll—just tell us when."

I promised to get back to them as soon as I put the pieces together. The next call was to Ray Petersen, president of Wien Consolidated Airlines, to ask for another favor.

"What is it? Anything but cash," Ray said. We both hooted.

"Ray, I would like your flight to Saint Mary's to remain on the ground for an extra two to three hours this coming Thursday."

"I don't know about that," he said, sounding skeptical. "Why is it you need the departure delayed so long?"

"Ray, I'm trying to get a nun in Saint Mary's her U.S. citizenship.

Judge Ray Plummer has agreed to convene court in the village so this woman can take her oath. I think we can do this in two hours."

"I'll get right back to you," said Ray. "I will call the crew and see if they will cooperate. But I'm fine with it. I'll bet I know her. What's her name?"

"Her name is Thecla; she's from Italy. And I just found out there's another nun there, too, who wants to become a citizen. Her name is Sister Scholastica Lohagen, who was born in Germany and has been at the mission for more than forty years."

Within hours, everything was coming together. But I had another crucial call to make, this one to Senator Gravel in Washington, D.C. Mike and I had served together in the Alaska legislature.

"Hey, Mike I need a big one," I told him.

"Let me guess," Senator Gravel said. "You want an airline route direct from Paris to Nome?"

"Nothing that easy," I said. "I just want you to steamroll some paperwork through the federal bureaucracy."

I told him that between them, two elderly nuns at the Saint Mary's mission had waited more than ninety years for their American citizenship and needed help getting it done. "I've got their applications filled out," I said. "I even signed their names for them. I'm going to Express Mail the applications today. Can you can get them approved in a day?"

The senator promised to have a member of his staff hand-carry the applications to the Immigration and Naturalization Service and wait for approval. "I'll instruct them to stand there with their hands out twenty-four hours a day if need be," he said.

The following day, Senator Gravel was on the phone. "The sisters' applications have been approved and the paperwork is in the mail to you. Good luck."

On Thursday morning, the Wien flight to Saint Mary's carried an eager and excited group of passengers and crew. The first surprise of the day awaited as we touched down on the gravel runway in the Fairchild F-27 turboprop. We were greeted by at least fifty people, happy and waving.

The captain shut down the engines, and we piled out. The crowd of mostly Eskimos dressed in fine furs was there to escort us into the village. They insisted that the judge, his clerk, the flight crew, and I climb onto the back of an old Army surplus truck and ride to the school—a distance of a few hundred yards. This old truck was the only motor vehicle in Saint Mary's.

Another surprise: the gymnasium at the school was packed. Many of the people couldn't speak English, but everyone knew exactly what was about to happen. Many had traveled by dog team and snow machine. A few had even walked a hundred miles to Saint Mary's from tiny villages scattered across the tundra. Everyone wanted to witness this important ceremony.

We looked at one another in surprise. I could see the emotion in the faces of Judge Plummer and members of the flight crew.

Ray Plummer emerged from a dressing room in a black robe, representing the United States of America. A table flanked by the stars and stripes and the Alaska state flag stood ready in the center of the basketball court. Plummer took a seat. The clerk said in a booming voice, "Hear ye, hear ye, the court of the United States of America is now in session."

Dressed in their black habits, the two elderly sisters faced the judge and solemnly raised their gnarled, work-worn hands.

The audience seemed to hold its collective breath as Sister Thecla, age eighty, and Sister Scholastica, seventy-five, pledged to defend their country against all aggression, foreign or domestic. Then everyone joined in singing "God Bless America."

There wasn't a dry eye in the place.

It was Christmas Eve, December 24, 1969.

Chapter 19
Sixty-eight Below Zero

Anchorage, 1970

My latest acquisition was a 250 horsepower Piper P-24 Comanche. Charlie accompanied me to pick it up in McAllen, Texas. From there, we flew out over the Gulf of Mexico and swung around to Port Richie on Florida's Gulf Coast about fifty miles north of Saint Petersburg. My parents had moved to this area. After looking around a bit, I landed in an unfinished subdivision right on the beach. My dad met us there.

A county sheriff's deputy showed up as we were throwing our bags into the back of Dad's pickup truck.

"What are you doing here?" he asked.

"I just stopped by to visit my folks. They live right over there," I said, pointing.

"I mean, why land here, instead of at the airport?" he asked.

"I want to visit my parents, not an airport twenty miles away," I said.

The deputy scratched his head trying to figure out what kind of ticket to write.

"Officer, there are no federal air regulations about where you can land an airplane. I haven't hurt anything, and I'll be gone tomorrow morning."

The deputy accepted this and drove away.

The next day, I inspected the airplane's mud-encrusted landing gear, the mud having hardened like concrete, and decided it would be prudent to leave the gear down when I flew to our next stop. Mom and Dad would meet us at the Albert Whitted Airport in downtown Saint Petersburg, where I would refuel and wash the undercarriage. As we rolled up to the office of the local flying school after the short flight, I spotted just what I needed—a garden hose.

I got out, uncoiled the hose, and crawled under the wing, spraying the dried mud that covered the landing gear. My mother and the manager of the municipal airport both showed up at the same time, and he began to berate me for my off-airport landing the previous day. My mother spoke up before I could say anything.

"Look," she said, "this is what my son does day in and day out, year-round—land in rough places including sand bars, hilltops, river bottoms, mud flats, even dirt roads when one is available."

I could have hugged her on the spot. I kept my mouth shut, washing away, as she backed my critic into the sanctuary of his office, turning the conversation into a lecture on the care of retractable landing gear. After coiling up the hose, I squeezed every drop of fuel I could into the tanks, and we took off for the sunny climes of Alaska 5,000 miles and four time zones away.

After an overnight stop in Cheyenne, Wyoming, the flight home was routine until we closed in on Whitehorse, Yukon Territory, where the temperature was dropping fast. Whitehorse reported thirty degrees below zero with ice fog, so we stayed up at 10,000 feet where visibility was unlimited, and filed a flight plan to Northway, Alaska, on the U.S.-Canada border. I checked in by radio as we neared Northway. How were we doing in the cold, the FAA wanted to know.

"Not too bad, my outside air-temperature gauge indicates minus thirty-five degrees but all systems seem to be working," I replied.

We were told Northway was experiencing ice fog, with greatly reduced visibility, and the current temperature was sixty-eight below.

"We will turn the runway lights to full intensity," the FAA said.

I began looking for the runway while trying to work out in my mind how to deal with the extreme cold. I figured my best bet was to lower the landing gear before beginning our descent, while keeping the fuel mixture lean and some power applied. It was crucial that I keep the old Lycoming engine running.

Circling the airport, I decided on final approach I would touch down as short as possible on the runway to avoid having to use the brakes. Likely the metal parts would have contracted in the cold, leaking fluid, and it was doubtful the brakes would do us any good.

Fortunately, I had plenty of room. The 5,100-foot runway had been built to accommodate Allied bombers and P-39 Airacobra fighters being ferried to Russia as part of the Lend-Lease program during World War II.

The landing was smooth. Coming to a stop next to the fuel pump, I adjusted the throttle to about one-quarter and placed the mixture control on rich so I would be ready for a quick restart. Then Charlie and I jumped out. Charlie headed into the restaurant for hamburgers and fries to go. I wrapped a big Eddie Bauer down jacket around the prop hub, stuffing the sleeves into the nose of the cowling to preserve a little engine heat, and went to clear U.S. Customs.

After my son and I finished our business, we jumped back into the Comanche, glancing at one other nervously as I hit the starter button. The prop moved slowly, but caught, and the engine roared to life. What a relief. I looked up to see a gang of onlookers clapping their hands in front of the Customs office.

* * *

DURING MY TIME at Christiansen Air Service in Bethel, having acquired half interest in the operation, I purchased three new Cessna 172s and a succession of Cessna 206s and 207s. They came from the factory equipped with "wheel pants," as they were called, to reduce drag and improve the appearance of the aircraft.

As soon as we tied down a new airplane, we would go out and jump up and down on the plastic wheel pants, breaking them off. Then we threw the parts onto a junk pile and set about replacing the stock tires with oversized tires better suited for Alaska conditions. The wheel pants and other plastic parts were worthless in subzero temperatures.

I tried different methods to keep aircraft engines and the oil in them warm enough in cold weather to allow the oil to be drained and brought inside at the end of a day of flying—always a good precaution against freezing.

For years Standard Oil Company and Union Oil sold aviation gasoline in square, five-gallon cans, packed two to a wooden box. It seemed as if everyone in Alaska and the Yukon had converted the empty cans into water cans, cooking pots, and who knows what else, just as used tin coffee cans became light fixtures and other durable fixtures.

I once made an engine heater out of a couple of fuel cans and a simple plumber's pot. It worked great, required no electricity or fancy fuel, and in a pinch you could build a small wood fire in it. But you had to be careful. A few pilots burned up their planes due to inattention.

One January I was on my way from Florida to Alaska in a Cessna 170A when I stopped for the night in Fargo, North Dakota. I checked into a Holiday Inn and went to bed. During the night, I was awakened by a racket out in the hallway. People were scattered all over the place curled up on the floor asleep or trying to get comfortable. A dangerous ice storm had swept across the northern plains, closing the main east-west highway. To its credit, the Holiday Inn threw open its doors, giving shelter to one and all.

When the weather began to improve, I headed out to the airport to see how soon I might be able to get under way. Six aircraft were waiting in line to use the only big space heater available. The pilots and ground crews were directing the flow of hot air onto the tops of the wings, slowly melting the ice covering.

"Guys, can I use this thing for a few minutes," I asked, stepping to the head of the line. "I think I know how we can move this along faster."

"Yah, give him a try—anything is better than this," somebody said.

I found a screwdriver and removed the inspection covers from the underside of the Cessna's wings. Next, I stuck the air hose up against one of the openings. In a moment or two, hot air inside the wing had melted a thin layer of the ice on the surface and a large sheet of ice slid off on the ground.

A cheer went up.

* * *

BY 1971, I HAD burned out on politics after losing a race for lieutenant governor. Visions of warm ocean beaches and palm trees beckoned. Ann and I kicked around ideas for a couple of days and I decided to move back to the Lower 48. We thought Florida was a good choice. It has lots of airports, we found some good schools, and I could make a living selling airplanes anywhere. We settled in Plant City and sent the boys to a big new high school there.

On my way south in a Cessna 180, I had spotted a good buy in Seattle. So, before I could unpack, I was on the way west again to pick up a 1946 Fairchild F-24 with a 165-horsepower Warner engine. Dad came along for the ride. From Seattle, we followed the West Coast south as far as Santa Barbara, California, then headed east across country to Lakeland, Florida.

The first time I took off at night in the Fairchild I thought something was wrong. The exhaust-collector ring was leaking sparks, flames, and clinkers. Once it got hot, it swelled up and sealed itself. It had been doing this all along, I just couldn't see it during the daytime.

The single-engine four-seater cruised at about the same speed as a new Cessna 172. It was an old-fashioned rag and tube tail-dragger you could fly with your elbow stuck out the roll-down glass windows. The Fairchild was a beautiful plane with nice manners.

I sold the Fairchild right away and headed back out to New Mexico where I bought the first of four Cessna 195s. I flew the plane from Las Cruces to Lakeland with no right brake. I stopped at three different shops on the way looking for a fix. We put in several quarts of fluid. But the brake did not come to life.

So, to make a right turn on or off a taxiway, I had to make a 270-degree turn to the left. This baffled a few onlookers. Within a week or so, I had sold the 195 and bought my first Globe Swift near Atlanta. It was bent in a few places, having been taxied into a hole on a grass field. My dad and I loaded it onto his boat trailer and hauled it back to Port Richey. With parts flown out from Colorado, we had the plane back in the air in a few days.

The plane was like an eager sled dog, ready to go when I got to work in the morning. I flew all over the South looking for good buys, especially single- and twin-engine planes, which would turn quickly. I was selling three or four planes a month.

I added a sideline of ferrying aircraft—anywhere, anything, including some flying wrecks. Many jobs came mostly from insurance companies. Once I flew a Cessna 320 from the Virgin Islands to West Palm Beach, Florida, with a wing spar bent and the landing gear wired down and pieces of two-by-four supporting the struts. I flew this mess over water for seven-hundred miles at about eighty-five miles an hour, dodging every cloud on the way.

Later, I got a call from the Aircraft Owners and Pilots Association in Washington, DC. A member needed a Cessna 185 on floats to be ferried from Minneapolis south to the headwaters of the Amazon River. The owner planned to haul tropical fish from the upper Amazon to Barranquilla, Colombia. This sounded like an interesting job, so we negotiated a fee.

I ordered the necessary charts and set out to secure whatever permits would be needed. Then I waited. Finally, not hearing anything for a month, I called the aircraft owner and asked about a departure date. He told me he'd found a flight instructor looking to build his hours who offered to ferry the plane for free. The 185 owner elected to go the cheap route. The flight instructor crashed in the Magdalena River near Barranquilla. The plane was totaled, as was the flight instructor.

Within a few years, I was ready to get back to Alaska, where I could earn a living flying light planes. My wife, two sons, our dog, and I climbed into my Cessna 180 and headed home. It was summer, the fish were running, and Japanese buyers wanted as much salmon roe as they could get their hands on. The Cessna 180 on floats earned me more money in two months than I'd made in Florida in a year.

But I'm getting ahead of my story. About this time, just as we were packing up and ready to go, Earthquake John McBride shows up at the front door, his face full of smile. He was wearing his usual flying uniform, a food-spattered jumpsuit and slippers. I was glad to see him.

John said he was flying a Lockheed Constellation for a cheapo freight operation in Miami and needed a copilot. I figured, why not? It paid fifteen dollars an hour. I could do this for a while.

On our first trip, we took off from Miami heading for San Juan, Puerto Rico. As soon as we were airborne, and I was established on our route, John pulled back his seat and went to sleep, mouth wide open, snoring away. As we rumbled along over the ocean, I got a closer look at the panel and radios. The autopilot didn't work. Most instruments were swinging back and forth on wires, or were simply dead.

It didn't take long to figure out why I had been hired. It was cheaper to hire me than to fix the instruments. I had a reputation for flying on and on for hours, the ultimate cheap autopilot.

I was on short final at San Juan when I nudged John and asked him if he wanted to watch the landing, my first in a Connie. He groaned and reluctantly opened an eye.

Soon we were hauling baseballs to Haiti, where covers were hand-sewn on. Then mangoes back to New York. We also hauled a lot of live white Charolais cattle to San Jose, Costa Rica, and boxes of fresh beef from La Ceiba, Honduras, back to the States.

I was getting the hang of flying those big piston-pounders but their days were numbered. Most of them needed hundred-forty-five octane fuel, which was in dwindling supply. On our way back from Central America, we would stop at Isla San Andrés, an island in the middle of the Caribbean belonging to Colombia where we filled our tank with hot fuel, as the high-octane fuel was called.

The planes were worn out, the engines shot. We would go out on four engines and return on three. Ann was getting nervous.

John called one day to say he had an interesting job with lots of hours. We were to fly a load of office equipment from New York to Milan, Italy, and then pick up some racehorses in the Balearic Islands off the coast of Spain and take them to Nicaragua. The horses belonged to Anastasio Somoza Garcia, the long-time Nicaraguan dictator.

"No, thanks," I said. "I don't think there is an airplane in the bunch sitting in pools of oil at the northwest corner of Miami International Airport that could make the trip. It's not called 'corrosion corner' for nothing."

John was gone for a month. He lost an engine in the middle of the Atlantic and spent a week in the Azores feeding and watering six racehorses. Fortunately the plane had built-in bunks because none of the three crew members had any cash for expenses. They were jumping for joy when a plane arrived from Miami, but were crushed to learn it did not carry a spare engine and mechanics to install it. They never got paid for the last leg of their trip on three engines.

The next time John asked me to fly with him on a delivery, I climbed into my Spitfire—really a Swift painted to look like one, complete with British markings—and flew to Fort Lauderdale. John and I drove over to look at a ratty-looking Lockheed Constellation. I didn't like this airplane. My reaction was strange because up to this point I had always been ready to marry almost anything built to fly. Flat tires and leaking hydraulic fluid and oil be damned, I *always* had found something of beauty under the makeup. But not this time.

I crawled over her and then studied the logbooks. There had not been an entry in five years.

"John," I said, "I've had it with this."

I didn't stop ferrying airplanes altogether, but I did become more selective. And I didn't quit buying and selling planes, either. A few years later, back in Anchorage, I became the dealer for Piper Aircraft Company. The routine was predictable and got me plenty of flying time. I would fly commercial to Florida on Friday afternoon, pick up from the company a new Seneca Twin PA-34 or a Cherokee 6, leave Florida on Saturday morning, and be back at Merrill Field on Sunday afternoon. That gave me about twenty-six hours at the controls over two days.

Chapter 20
The Climber

We knew that Charlie and his two climbing companions, Brian Canard and Robert Frank, would be nearing the summit of their climb. It was a beautiful clear day, the kind you live for in Alaska.

I decided that we would take the Piper Turbo Arrow, a four place fast little aircraft. My son Richard took a back seat and Ann and I climbed in the front. As we departed Merrill Field and began our climb I did the math. They had been on the mountain for about a week so they should be nearing the summit. Mount Deborah at 13,000 feet or so was not a giant like Denali but it was a highly regarded and difficult climb.

I had told Charlie some years back to be careful in his choice of pilots who would be flying him in to the base camps in Alaska. "Charlie, always make sure the pilot has grey hair." This time, I knew he was in good hands. My good and trusted friend Cliff Hudson from Talkeetna had flown him in to Deborah.

This was early in Charlie's climbing career, but even then FedEx trucks were delivering all kinds of climbing gear to his house every day. The manufacturers of tents, boots, climbing tools and clothing kept a steady stream of products coming his way. He would try them out and give them a thumbs up or down. He was a sponsored athlete. I wasn't doing too badly either. Piper Aircraft company through their distributor Melridge Aviation in Vancouver, Washington made sure that I had the latest and best aircraft they produced. They just handed me the keys along with a credit card for fuel and said, "Go get 'em; sell this thing." As a result I had a new six place Seneca 111 or a 10 passenger Navajo Chieftain for my exclusive use. Not bad for a guy who loved airplanes.

Charlie had shown me the route they would climb on a map. It was clear from studying the map that this would be a very risky climb. The mountain was oriented East and West. They would be attempting the

West Face, mostly climbing on snow and ice. The route would mount a ridge angling steeply towards the east.

I knew that Ann was a bit apprehensive. We rarely had an opportunity to see our son actually on a climb. This was one of those times when I knew from the weather that we would stand a good chance of seeing him doing what he loved doing; he was one of the world's top high altitude mountain climbers.

After I passed Denali off to our West, I could see Deborah straight ahead; it was a rock spire standing alone. The air was absolutely still. There was no cloud cover and the visibility was better than 100 miles - perfect for our approach to the mountain. I closed in at about 12,000 feet where I thought Charlie and his two partners might be.

There they were, with Charlie in the lead with 100 feet of rope between them, climbing the knife edge at about a 75° angle. The wind had blown the snow over the edge of the precipice so it was sticking out over space. Like a spring board.

I reduced the power so we could slow down and pass very close to Charlie. My concern was that the noise of the engine and propeller, or the downwash of air coming off the plane would cause the snow to break off under them. As we coasted by them, they looked over at us but didn't even wave!!

I saw why when we realized they were standing on a snow cornice about six feet wide. Ann let out a gasp, and Richard said, "Oh shit!!!" over and over, and over. There was about 6,000 feet of space between Charlie's feet and the next piece of rock, straight down the sheared off North Face. Absolutely unbelievable.

The Great Wallenda walking a high wire across Niagara Falls was a stroll in the park compared to this. I looked at Ann's face, which as usual had a look of concern, but not panic. This after all was her firstborn child.

It was like he was standing on a single bed mattress which had been placed on top of five Empire State Buildings. I had to hand it to him, he sure knew how to impress his parents.

I took a wide swing out away from the mountain and circled in for a second look. This time as we passed by I noticed he was moving his left hand which was by his hip, up and down a few inches in a mini wave. I knew then that Charlie was fully aware of how precarious their situation was, even though he couldn't see out and take a peek over the edge.

Within a few hours the top of the mountain was enveloped in a raging wind storm. Charlie told me later that they were forced to dig a cave down into the hard packed snow where they huddled for three days listening to the howl of the wind and the creaking and cracking noise of the walls of their shelter.

When they climbed out of Cliff Hudson's airplane five days later in Talkeetna, we were shocked. All three of them appeared to have aged 20 years.

A couple of years later when Robert Frank, Charlie and two other climbers did the first winter ascent of Mt. McKinley's West Rib and the third winter ascent of the mountain, they went by way of a 9,500 foot ice rib and face. It was in February; there was no one else anywhere in the National Park.

Steve Teller was injured when he slammed into the ice during a fall on the route and Chris Hrayback had run out of energy. They were lashed to the face of the mountain at about 17,000 feet. Charlie and Robert continued up the mountain and made the summit. They sat down right at the top, 20,320 feet high, and dug out a bag of cookies that Ann had made for them.

After enjoying the view for about 20 minutes they began the descent. Down they went unroped because under these conditions there is no way that a lone climber can arrest the fall of another. The ice was so hard that crampons didn't even make a scratch, Charlie was in the lead. Suddenly Robert Frank fell and slammed into Charlie. Down they tumbled. Charlie was on his back sliding down a 60 degree slope at 18,500 feet. He managed to roll over like a cat and get his ice ax under him and arrest his slide. A quick glance over his shoulder confirmed the worst. Robert Frank was tumbling out of control never to be seen again.

Charlie was left in a very precarious situation with one crampon off dangling from a cord and the ice below him hard as concrete. He managed to stand up on one foot and chip a little spot to wedge a toe in while he reinstalled the loose crampon. He then climbed down to the other climbers, who were hanging in sleeping bags lashed to the ice.

When Charlie told them what had happened, they said "We didn't see him but we heard him as he went by us." Charlie continued to follow Robert's fall line, finding his down jacket, which had been flung off as he rocketed end over end, hitting the wall only occasionally. He also found a 35mm camera body, and some blood. Apparently Robert's body finally found a resting place among a gathering of big ice chunks at about 13,000 feet.

The realities of the situation required Charlie to contain his grief and go about the business of rescuing the other two climbers. He climbed back up to them and told them what he had found.

With their injuries and somewhat limited energy he told them there are two ways down. "We can either go up to the usual descending route off the mountain that will take us two to three days to reach the base camp which is 6,000 feet directly below us. Alternatively we can rappel straight down from here."

"We should be able to make that in less than a day. The problem is we don't have enough hardware to make up the rappel points. Between us we only have 15 pieces of hardware, and I believe we're looking at 23 rappels. We only have enough stove fuel to make water for one day. If we go up and around by the traditional route, we will probably die. On the other hand if we rappel down I'm going to have to find natural rappel points or chop them into the ice, so there's a good chance we could die there too. What do you guys want to do?"

Without hesitation they both said, "we will follow you." Charlie looked them in the eye. "Then let's get moving. We are going to rappel off. Each of you check your gear carefully before each move. No one including me is going to lean backward until it's clear we are properly hooked up.

Ann and Charlie's girlfriend at the time and I had decided to take a look at the mountain because we thought they would be down to their base camp at 10,000 feet. We flew in to the mountain at about 15,000 feet. We were flying a new Piper Seneca six passenger twin-engine plane that the Piper aircraft company had given me as a demonstrator.

We flew around where we thought they would be, but could never locate them. So we turned around and went back to Anchorage. The following day we tried again this time in a Cessna 185. I thought the 185 would be better since we could maneuver in tighter quarters. This time as we circled down in an ever tightening circle in the cul-de-sac, we saw them.

Their base camp, which consisted of one small tent, was sitting in the dead center of the tiny snowfield, right at the bottom of the slope they had just climbed. We saw them standing by the tent. We couldn't get close because of the confinement created by the encircling ice falls. We all gave a sigh of relief however, knowing that they were down and safe.

They would have to break camp and ski down to 7,500 feet where a plane would pick them up. The next day we flew to Talkeetna and waited for them to be brought out. When the airplane touched down on the city strip and slid up to the Fairview Inn, only three climbers got out.

As Charlie walked toward me, I could tell something was wrong. "Dad," he said, "Robert is gone." I put my arm around him. There aren't real words that describe sorrow in these situations. Only a tiny group of human beings have ever reached the outer limits of the globe we share. Charlie continued to climb in Alaska, Europe, South America and the Himalayas. As of this writing he is the president of the American Alpine Club, a worldwide organization.

Chapter 21

James Irwin

Cape Canaveral, Florida, 1971

My dad phoned with exciting news: my cousin James Irwin was going to fly to the moon the following week from Cape Canaveral. Ann and I were in Florida at the time.

"We should use my boat to get up close," Dad said. "There will be a big mob watching the launch."

My mother, Ann, and the boys were with me when we launched Dad's boat into the Banana River and worked our way up close to the launching pad. What a thrill it was to see and hear that big rocket go.

Jim and I were almost the same age. He had attended the U.S. Naval Academy while I was at UCLA. I recall we both made our first solo flights in 1951. While I was flying in Alaska, my cousin had been advancing through the military system as a test pilot and later an astronaut.

Jim had taken special classes in geology at Caltech before flying to the moon on the Apollo 15 mission, the fourth in a series of manned moon landings. NASA had wanted him to know what he was looking at when he got to the lunar surface.

He had a specific mission and prescribed route in the Lunar Roving Vehicle (LRV) his first day on the Moon in August, 1971. He spotted an interesting piece of rock, but due to the tight schedule, was unable to stop to examine it. The next day, however, Jim drove back for another look and brought the rock back to Earth.

My cousin had hit the jackpot. When NASA's astrogeologists analyzed the sample, which came to be known as the Genesis Rock, it was found to be more than four billion years old. To this day, scientists say, it continues to yield clues as to how the solar system was formed.

As an Apollo astronaut, Jim spent more than two-hundred ninety-five hours in space, and more than eighteen hours on the surface of the Moon. Not bad for a boy who got his start building model airplanes.

Chapter 22
White-knuckle Landing

Lakeland, Florida 1974

Earthquake McBride was excited the next time he called. "Johnson Flying Service out in Missoula, Montana, has a good Curtiss C-46 they want to sell," he said. "Let's go get it."

The C-46 was the largest twin-engine tail-dragger ever built. It weighed 51,000 pounds and carried 1,400 gallons of fuel. It ran on hundred-octane aviation gasoline, which was easier to find than one-forty-five. This prospect sounded good, so John and I headed west in my 1948 rag-winged Cessna 170.

I was impressed with the C-46. The owners had planned to convert it into a spray plane, but changed their minds. It came with six spare 2,000-horsepower Pratt & Whitney engines, some needing reconditioning and others good for parts only.

I bought it for $30,000. Johnson Flying Service signed off a ferry permit so we could fly the aircraft back to Lakeland, Florida, where I was operating at the time. We loaded the four best engines, chaining them to the deck. They weighed 2,500 pounds each, but we were still under gross weight.

I climbed into the left seat while John called off the checklist. The brakes squeaked and groaned as we taxied out onto the runway. Getting a nod from John, I opened the throttles to full power and we rolled down the runway. I marveled at the good view with the nose stuck up in the air, tail wheel still on the ground. Up to this point, the only tail-dragger twin I'd flown was a Cessna T50/UC78 "Bamboo Bomber," a postwar surplus aircraft with a pair of 300-horsepower engines. It was thrilling to have a grip on a pair of 2,000-horsepower engines.

Later I got a kick out of flying an empty C-46, too. Without a lot of weight this aircraft leaped off the ground with instant power from its propellers.

As we closed in on Lakeland at 1,500 feet, the approach altitude for heavy aircraft, a restored, twin-engine bomber suddenly cut in front of us. The bomber had come so close that I considered this a near hit, not a near miss.

When a collision seemed imminent, I immediately pulled back on the control wheel as hard as I could, and the engines began to misfire and shake, on the verge of quitting. My second natural reaction was to shove the wheel forward to regain flying speed, then turn on the electric fuel-booster pumps.

The C-46 has six fuel tanks, three per wing. Trying to increase fuel pressure took a few seconds. One engine was shut down; the other was giving us about twenty-five percent of full power. Meanwhile, we had descended to 1,200 feet. The ground was coming up fast.

"Pick up the landing gear," I shouted at John. We were going to hit in a short, soft field and I didn't want to roll. I just wanted to get on the ground in one piece and stop.

John was frozen, saying over and over, "We're going in, we're going in!" I reached down and pulled up the landing gear. Nothing left to do but make the best of it.

I summoned my piloting skills as I held the control wheel by my fingertips, as lightly as possible, feeling for the tiny shudder that would indicate the C-46 had reached the edge of a stall—a condition in which the airplane becomes an uncontrollable paper weight, usually spinning down to Earth and killing everyone on board. This is why airplane wreckage so often shows a tail sticking out of the ground.

We were in a controlled dive with the nose of the C-46 aimed for a spot in the farmer's field. I kept my eyes fixed on the spot and literally walked the plane down by the feel of the feedback from the elevators in my fingers. Time slows down for many pilots in these kinds of situations, and luckily I was one of them. I followed the touchdown spot all the way down, almost to a point at which the nose was close to smashing into the ground, then hauled back on the wheel with everything I had. This pulled the diving plane into a nose-high altitude, and it did a gigantic belly flop just as it stalled out.

A cloud of dust, weeds, and chicken feathers filled the air as we slid about four-hundred-fifty feet to a stop. It wasn't a pretty landing, but it worked.

I opened the cockpit emergency door, unzipped my pants, and let go a stream of pee. People were running across the field toward us, but I didn't care.

John was speechless. I felt sorry for him. Thousands of hours flying big transports hadn't helped him when he needed it.

A National Transportation Safety Board (NTSB) investigator was standing by the plane when I arrived the following morning.

"Who was flying this plane when it landed?" he asked.

"I was," I said. "I'm the owner."

"Let's see your license," he said. I handed him my license, which read "commercial pilot, single engine land and sea, multi engine land, instrument."

"Where is your type rating for a C-46?" he asked.

"I'm going to get one as soon as I learn how to fly it," I answered.

A "type rating" is required for aircraft weighing more than 12,500 pounds gross weight. He asked me how the fuel system worked, how to extend the landing gear manually, including the tail wheel, which is retractable on the C-46. I survived the NTSB's scrutiny, so no violations were filed against me. The pilot of the bomber that cut in front of us was not cited. But the agency did nail the mechanic at Johnson Flying Service, who had not cleaned the fuel filters as part of the previous owner's annual inspection prior to selling me the plane. This was the reason the engines died.

Chapter 23
A Wild Ride

Dominican Republic, 1975

My next task was finding the parts we would need to put the C-46 back in the air—a propeller, two wheel-well doors, two oil coolers, a bottom cowling, and about twenty feet of belly.

John McBride suggested I contact Dominacana Airlines, which operated C-46s out of the Dominican Republic.

"Chuck, they have four of them sitting way back in the weeds," he said. "I doubt they will ever try to put them back into flying condition. So you can probably buy what you need for next to nothing."

I gathered up my dad and Bob Wilcox, a good friend who had loaned me some money to buy the C-46. Bob was a radiologist who loved airplanes and took part in occasional adventures hanging out with the Sassara clan.

The three of us climbed into my Cessna 310 in Florida and flew across the Caribbean, heading first to Haiti, where I was negotiating a deal to fly mangoes up to New York when I got the C-46 into the air again. We spent the night in Haiti, drinking Barbancourt rum and Cokes until the wee hours, then headed over to La Romana International Airport in the Dominican Republic. There, we got caught up in the prolonged haggling that Latin Americans seem to enjoy in deal-making.

"Come on, let's get a move on," I urged my passengers after finally wrapping up the negotiations to buy spare parts. It was getting late in the afternoon, and we were returning to Haiti for the night. The timing was going to be tight because the airport in Port-au-Prince closed at sunset.

We took off heading southeast. As I turned onto the proper course bound for Haiti, flying at about 6,000 feet, we found ourselves in a heavy tropical rainstorm. Suddenly and unexpectedly I was flying on instruments as lightning flashed around us.

It looked as if we were in for an exciting ride. The vertical speed indicator was pegged to the top, indicating a vertical climb of more than

2,000 feet per minute, even with both engines at an idle. The Cessna's nose was pointed down at about a 30-degree angle as the wind sucked us ever higher. The giant thunderheads were formed around a vertical column of air rushing upward in the middle, pouring out the top, and then rushing downward on the outside of the column. Radar-equipped planes see these conditions and avoid them, but I had no radar, and there was no turning back.

Because the giant thunderhead was embedded within the clouds, we were not aware of it until conditions became so severe my main concern was holding the plane together in one piece.

Thunderheads in this part of the world can boil up to over 50,000 feet. Holy Nellie! I was caught in a giant vacuum cleaner. These things can digest an airliner. Turning in any direction was out of the question. I watched the altimeter as we passed through 13,000 feet, still without power. I glanced over at Bob Wilcox in the front passenger seat. He had a death grip on his seat and was staring straight ahead. When I suggested he call Port-au-Prince for me, he just shook his head no. I wondered how Dad was doing in the back seat, but I dared not turn my head around to look.

Then the rate of climb began to slow. I eased in a little power and leveled out at 17,000 feet. As I turned the plane to get on the airway I was confronted with another problem: no radio aids were working. Neither the VOR nor ADF indicators showed any sign of life.

I picked up the radio microphone. "Port-au-Prince radio, Port-au-Prince radio, this is Cessna N31GB." I checked my watch and calculated our position. I estimated we were nearly clear of the 10,000-foot mountains on our route and would arrive in the Haiti capital in about thirty-five minutes.

A few more radio calls confirmed I wasn't going to get any help from the ground. The airport had closed for the day. I knew that Haiti did not allow flights in or out after dark, but I figured I might be forgiven a late landing since I was on an international flight plan and flying in risky weather.

The thought of returning to Dominican Republic with that thundercloud in the way was out of the question. Jamaica was a possibility, but it was a long way and I did not have any approach charts. So I kept chugging along on my compass course while running through a few other options. Guantánamo Bay, Cuba, maybe? Surely that would raise a few hackles.

When I had reached a point where I was certain that I was out over the water west of Haiti, I started my let down. It was pitch black, still raining heavily, and we were surrounded by thick clouds. It gets dark at six o'clock in this part of the world. There is no twilight.

I was doing two hundred miles an hour at about eight hundred fifty feet in solid IFR conditions, heading for where I thought I would find Port-au-Prince. I kept looking up from the instrument panel, hoping to see lights on the ground. If I did not see shore lights within five minutes, I was going to turn around, climb to at least 5,000 feet, and try a new approach.

Suddenly there was a glow straight ahead. As we closed in on the shoreline, a swatch of city lights began to appear, except for a large hole of darkness in the middle. This had to be the airport, without a damn light anywhere. No runway, taxiway, street, parking lot, or building lights—a complete black void.

By then I had descended to four-hundred feet, coming in fast. It was time to slow down, drop the gear, and find the runway. This might require some slow S turns until I became oriented, but after that it would be easy—or so I thought.

Suddenly two bright white lights came on directly ahead of us. They were lights on an airliner running down the middle of the runway, taking off in our direction.

I shut off my navigation lights so the pilot of the airliner wouldn't get spooked and abort his takeoff. I continued my approach figuring I'd pull a hard left turn if the jetliner didn't lift off and get out of my way. This was going to be close. I continued on in, passing under him by a couple hundred feet. Crisis averted. The runway was straight ahead. I turned the lights back on and landed on a lovely stretch of pavement. I thanked my lucky stars I had thousands of hours of flying multi-engine aircraft with sectional charts open in my lap. They had paid off again.

The next day I learned the jetliner was an Air France flight headed for Paris. Out on the runway the captain had delayed his takeoff due to a minor issue in the cockpit, and it became dark before he resolved the issue, turned on his lights, and blasted off.

Despite my sometimes unorthodox ways, I never had any trouble with the authorities in Haiti. This may have been due to the steady supply of Jack Daniels whiskey I presented to the head of the secret police. But

I have to confess, I was scared the first time I met him. His office was in the main terminal at the airport behind a large mirror, which turned out to be one-way glass. I was ushered in by a couple of big men packing large pistols. There he sat behind his desk, your worst nightmare. One look and I was ready to drop to my knees and plead for mercy, even though I was making a social call.

I found out that he had a taste for good whiskey, so the next time I landed in Port-au-Prince I presented him with a small desktop bar shaped like a barrel. It held a set of glasses and a fifth of Jack Daniels. After this became a fixture on his desk, I was treated like visiting royalty. I would roll to a stop in front of the terminal, shut down the engines, place a new soccer ball, a pair of sneakers and maybe a box of candy on the front seat for the children of the head of customs. Then I would breeze through customs smiling. Meanwhile three soldiers would take up position, guarding my plane twenty-four hours a day. Another team would polish the aircraft.

One day I was reflecting on this while relaxing on the veranda at the home of a successful attorney in Pétionville, an affluent suburb in the hills above the dust and clatter of downtown Port-au-Prince. This was the life: sipping my drink, enjoying the view, and smelling the flowers. Or was it? The problem was that I was losing money flying in this part of the world. The more I flew, the less I had in the bank. I thanked my friend and host, René, for his hospitality.

"I've got to get out of here," I told him. "I think I'll head back to Alaska tomorrow where I know how to earn my keep flying airplanes."

I called my dad in Port Richey and told him to pack his toothbrush. "Pappy," I said, "let's head for Anchorage—I'll pick you up in the morning."

A week later I was flying through Lake Clark Pass, dodging rocks and clouds on my way to Hooper Bay where I would deliver five Eskimo kids home from college. As I made my way through a light snowfall, I felt better than I had in months.

I was back in my element.

Chapter 24

Uncle Ted

Anchorage, 1975

Ted Stevens and I met on the campaign trail in 1964, both of us running for the Alaska Legislature. He had been a fellow pilot. His flying time had been brief but vital, taking Army C-46 transports over the "Hump"— the eastern end of the Himalayan Mountains—from India into China with arms and supplies to support China in its defense against Japan in World War II.

After the war, Ted went off to Harvard Law School, passing up an opportunity to become a commercial airline pilot. I often kidded Ted, questioning why anyone would give up a perfectly good aviation career to go into the law and politics.

I took a call at Christiansen Air in Bethel one day from the chairman of the newly created NTSB. He had been referred to me by Ted, who was appointed to the U.S. Senate in 1968 upon the death of E.L. "Bob" Bartlett and then elected to the seat two years later. The NTSB chairman wanted to learn more about flying in Alaska.

"Sure," I said, "what do you have in mind?"

"How about I fly up there?" the NTSB official said. "I want to hear your thoughts and suggestions about aviation in the north."

A few days later he showed up with an aide in a small government jet, a Sabreliner as I recall. I picked them up in the company van and did a drive-by to introduce them to our aircraft parked at the airport—a Cessna 402, an Aero Commander 560, a couple of 180s and a 207. Back at the office, alongside the river, I showed off our float-equipped Pilatus Porter, a 206, and two 185s.

Later, as I refueled one of the airplanes, I began to recite my laundry list of complaints, mostly having to do with lack of properly trained pilots. The pool of flying experience was small. But new pilots were having difficulty logging the flying time to meet federal requirements placed on

Alaska air-taxi companies in those days.

Lloyd's of London reserved the right to approve every pilot we hired before he would be named on our insurance policies. My company's minimum hiring requirement for conventional aircraft was 1,000 hours of Alaska flying time and 1,500 for floatplanes. The U.S. Department of Interior also required 1,000 hours of Alaska flight time to qualify for issuance of special licenses authorizing a pilot to fly federal employees in the state.

A steady stream of applications came across our desk from people who did not have enough experience. I felt bad turning away so many who had their hearts set on flying. Sometimes we heard from airline captains trying to get their sons into the business, thinking Alaska was a good place to start. Sorry, fellows, no can do, I told them. Some pilots even showed up unannounced on our doorstep with log books full of time spent as instructors. Their faces would drop when I told them flying around in an airport pattern was worthless experience for Alaska conditions.

For a while I considered starting an air taxi flying school, which I figured should require a commercial license for admittance. We would teach them to fly Cessna 185s, 207s, and C402s loaded up to gross weight with sandbags. There is a big difference between flying empty airplanes and flying planes fully loaded. The latter scenario, taking place in confined spaces, makes for a dicey go-around (a missed approach) in most light twin-engine aircraft. The go-around must be executed with precision, or you're dead. Yet this sort of crucial training is sadly lacking in most programs.

In my imaginary flying school, we would teach basic maintenance and troubleshooting—knowledge and skills needed to operate in the Bush. It is unfortunate that all the training in the world hasn't put a dent in the accident rate in general aviation. Most accidents are still caused by pilot error, and often it boils down to a single dumb decision.

In Bethel that day, I spent several hours with the NTSB chairman, who fired questions at me over coffee at the Kuskokwim Inn. I answered them honestly, noting situations in which my instructions to our pilots differed from FAA regulations. One of the issues was this: what should a pilot do if he found himself in rapidly deteriorating weather far from the sanctuary of an airport.

The problem was that pilots of single-engine air-taxi airplanes were not authorized to fly on instruments in Alaska, even when doing so offered the safest option in poor weather. I told our pilots with instrument ratings it was okay—no matter what the FAA said—to climb to a reasonable altitude and head for an airport with an approved approach. Then, to alert the nearest FAA flight service station. And finally, to land the plane safely. I promised to deal with any complaints.

Because of the inherent risks, I did not want the pilots flying low in bad weather. Our first responsibility was to protect our passengers. The chairman agreed with me and said to continue the safest practice.

Then he wanted to know about flying at night. "How do you land in the dark on those unlighted runways?" he asked.

I told him there were several ways to deal with this problem, and my favorite was kitty litter.

"What you talking about?" he asked.

I explained it was my custom to fill four empty coffee cans with kitty litter, pour in some diesel fuel, and take the cans out to villages I flew to regularly. There, I recruit a reliable local kid to position the cans at the four corners of the airstrip.

"When I buzz the village at night," I said, "the kid rides his snowmachine out, puts a little gasoline in the cans, and lights them. It works great."

The NTSB chairman later wrote me a thank-you note in which he said, "Don't change a thing, keep those passengers safe and, by the way, when I quit my government job I want to come to work for you."

Chapter 25
Heavy Lifting

Bethel, Alaska, 1977

There was one fairly constant trait among air-taxi operators who flew in the Alaska Bush—flying heavy loads. Nearly every takeoff was at gross weight. I am always reminded of this thinking back to the phone call one day in February from a purchasing agent at the Alaska Department of Regional Affairs.

"The village of Newtok is running out of diesel fuel for its generator. Can you give me a price per gallon and an idea how soon you could get moving on it?" the caller asked. Newtok was a village of several hundred people on the Ninglick River about a hundred miles west of Bethel. It had been an unusually cold winter, demand for electricity was up, and the village needed at least 4,000 more gallons of diesel.

This work was well within our capacity. We settled on a dollar-per-gallon surcharge on top of our standard hourly charter rate.

"Start as soon as possible and keep hauling as long as possible," I was told. Runway conditions at Newtok were ideal at the time, but could change depending on the weather.

"I'm practically in the air," I said.

We had priced the contract based on use of a Pilatus Porter, our biggest single-engine aircraft. A Swiss-built plane designed for short, rough fields, the Porter had been flown in Vietnam by Air America. This sturdy, rather slow airplane had a large cargo door, six removable seats, and a set of tracks in the tail. During the Vietnam War, the CIA sent people to Alaska looking to hire pilots for Air America. They offered me big bucks, but I was not interested.

"No, thanks," I told the recruiter, "I have a family to raise."

We loaded four drums into the Porter that morning in Bethel. The first trip went smoothly. I had thrown in two large truck tires to put under the cargo door for use as a landing mat. So far, so good; now back to

Bethel for the second load. As I lumbered along through the fading light it became clear the Pilatus Porter wouldn't do. It was too slow. No way could I get in two round-trips in the daylight available in late winter.

The next day, I rolled three drums into the back of the 206. This Cessna model is tough as hell, versatile, and capable of hauling five big passengers in and out of rough Bush airfields. It's easy to load, too, and performs well on floats.

I was able to make two trips a day in the 206. Bouncing to a stop on the Newtok strip, I rolled the truck tires into position and tipped the drums out the door. We established a routine: a couple of guys from the village would load a full drum onto a sled, drag it over to the generator shack, hand-pump the contents into the fuel tank, and then return the empties to the airstrip to be returned to Bethel for refilling.

By the fourth day, I had a good feel for the operation. The 206 was getting off okay with its heavy load and felt stable in the air. So I tried increasing the load from three to four drums. I was way overloaded but the balance felt right, so down the runway I went. The Cessna's powerful Continental I/O 520 engine was fitted with a two-bladed prop, which I liked better than the newer three-blade toothpicks. The two blades had a lot more metal on the leading edge to resist rock damage. When I got to Newtok, I set her down as softly as possible.

After a couple weeks of uneventful flights, two FAA inspectors showed up unexpectedly from Anchorage. They observed our operation and cried foul right away.

"Chuck, you are way overloaded," exclaimed Inspector Bob Labelle, who took out a pen and notebook and began scribbling.

"What you have here," Bob announced, "are 1,440 pounds of diesel and heavy-duty drums, 230 pounds of pilot, and 300 pounds of fuel for the plane. You are more than 600 pounds over the legal gross weight."

"Well, you are right about the weight," I replied, "but I can load her as heavy as I see fit as long as she flies okay."

"How do you figure that?" Bob asked.

"I'm under contract with the state of Alaska, which exempts me from FAA rules," I said.

Bob shook his head. "I'll check it out," he said, "but please, carry just three drums this morning, and I'll get back to you."

In the spirit of cooperation—Bob happened to be a personal friend—I unloaded one drum and took off. Bob returned in the afternoon.

"Our legal eagles confirm what you said," he said, "but we may require an airframe overhaul before you put this plane back into service under Part 135 of the air-taxi rules. So keep the weight down anyway."

I'll never reveal what happened next, but suffice it to say Newtok got through the winter with fuel to spare and I managed to stay out of trouble with the FAA.

Chapter 26
Robert Vanderpool

Anchorage, 1975-80

My friend Joe Vanderpool, who flew Fairchild F-27s for Wien, walked into my office at Merrill Field one day. Without prompting he helped himself to a cup of coffee and dropped into what I thought of as my "hot seat," a chair into which I maneuvered people who looked like prospective airplane-buyers. Joe didn't waste time getting to the point.

"You know my nephew Robert from out in Red Devil? He just flew into town to get his multi-engine rating, and the FAA won't give him a check-ride."

"Why the hell not?" I asked.

"They looked over his paperwork and logbooks and said everything looked good," Joe said, "but when the check-ride inspector went out to the ramp and looked at the aircraft Robert intended to fly, they said no way."

It seems the FAA had issues with a sixteen-year-old flying a ten-passenger Lockheed 10A Electra. The agency's Anchorage office claimed it didn't have anyone on staff qualified to fly it. Either that or they were afraid to get into it. The mental image was too good. I had to laugh.

"Bring Robert back here," I said, "He can use my Aero Commander 560. Surely someone over in the FAA will ride with him in this plane."

An hour later, Joe was back with his nephew in tow. The young pilot and I did a walk-around as I pointed to the nose, tail feathers, and wings of the Aero Commander, and then I showed him how to operate the throttles and props.

"Just bring her up to full power for takeoff and then reduce the RPM for climb," I told him. "When you level out, pull the props back to cruise RPM—it's that simple."

Robert fired up the twin engines and away they rolled, wandering from side to side, as the teenager figured out how to steer. I waved for him to stop and yelled over the engine noise, "Use your brakes until it gets rolling straight, then it's easy." They gave me a thumbs-up.

Two hours later Robert and his Uncle Joe were back, grinning like Cheshire cats. Ready for anything the FAA could throw at him, the teenager had taxied over to the FAA office and walked in with paperwork in hand.

"You back again?" somebody asked. Robert handed them the freshly signed application including the required recommendation endorsement from his uncle. Everything was perfectly legal.

"What kind of plane did you plan to take for the ride this time?"

Robert pointed to the Aero Commander.

"How much time have you got in it?" asked the FAA check pilot.

"About two hours," Robert said.

The man in the white shirt and necktie scratched his head, sighed, and said, "Okay, let's go."

An hour later young Robert Vanderpool had his multiengine rating. That afternoon, a Lockheed Electra loaded with supplies was rolling down the runway at Merrill Field. With 900 horsepower pulling him into the blue, young Robert started his flying career.

Two years later, at eighteen, Robert walked into my office at Christiansen Air on the edge of the Kuskokwim River in Bethel. He was all smiles. He had a new commercial pilot's license in his pocket and a lovely new bride at his side.

"Uncle Chuck," Robert said, "this is my wife Gail. I'm ready to go to work."

* * *

IN THE SUMMER of 1978, Robert Vanderpool called with bad news: his younger brother, Ronnie, had died in the crash of a floatplane in the Kuskokwim River at Red Devil, a small village named after a defunct mercury mine about one-hundred forty-five miles from Bethel.

I felt heartsick, knowing what the Vanderpool clan must be going through. Professional funeral services are not available in most Bush villages. The loved ones handle this themselves.

Ann and I decided to fly out to Red Devil for the funeral. For the two-hundred-fifty-mile trip over the Alaska Range, I picked out a four-place Piper Turbo Arrow tied down outside my office at Merrill Field. We pushed the Arrow to its maximum power across Cook Inlet, then over the mountains. A Cessna 206 was sitting on the strip at Red Devil. Bless the Vanderpools for anticipating we would need a lift to the patch of gravel behind their home in Georgetown, a settlement sixteen miles downriver. Our cab was waiting.

A crowd of mostly Athabascan Indians and gold-miners from up and down the river packed the house at Georgetown, where Ann and I were greeted warmly. The family insisted we sit up front next to the casket. The religious service was conducted in Russian by a Russian Orthodox priest. We didn't understand a word, but the meaning came through loud and clear: we were saying goodbye to a beloved nineteen-year-old whose life full of promise had been cut short.

After the service four Vanderpool men raised the casket to their shoulders and carried it outside, where a procession formed, everyone falling into line in twos and threes. We paraded around the house three times, many of the mourners sobbing. Our eyes welled up in tears.

Using ropes, the men gently lowered the casket into a freshly dug grave only thirty feet from the back door of the family home. As the grave was filled in with their bare hands, I was taken by the finality of it and by an overwhelming sense of love and human bonding among the survivors.

After the cross was installed and the last tear wiped away, mourners began to smile tentatively and murmur as we returned to the house where a funeral feast waited.

This is the way it should be done, with dignity and respect.

Chapter 27
The Two Hudsons

Talkeetna, Alaska 1981

Cliff Hudson was living in Talkeetna when we met. I had flown into the historic railroad and mining town one morning for coffee and a quick bowl of oatmeal.

The airstrip was unusually narrow but long by Alaska standards, about a thousand feet, and doubled as a street dead-ending in front of the old Fairview Inn. Pedestrians and motorists had to stay alert. The state was building a new airstrip just out of town.

As pilots invariably do, Cliff and I began talking about flying and then segued into a discussion of the challenges of flying in the vicinity of Mount McKinley, an imposing sight seventy miles away. At the time Cliff was flying a Piper PA-20 Pacer, but was thinking about getting into a Cessna 180 or maybe one of the new 260-horsepower 185s, either of which could pick up more weight. Don Sheldon, Cliff's neighbor on the airstrip, had bought a 180 and was poking around the mountain studying its possibilities.

Many landings later, after I began selling airplanes at Merrill Field, Cliff phoned to say he had heard on the pilots' grapevine that I had a turbocharged Cessna 206 for sale.

"I sure do," I told him. "It seems to be a good, clean airplane and hasn't been worked hard, having been a private plane all its life,"

I rattled off the pertinent details—the plane's flying time, engine, airframe, prop, and what I could remember about the radio.

"Sounds good, what do you want for it?" Cliff asked.

"How about $42,000, and I'll deliver it."

"No, I'll pick it up," Cliff said. "I need to do some shopping in Anchorage anyway. How about next Friday morning?"

"It'll be ready for you. I'll have the guys do a fresh hundred-hour inspection. Maybe you can pick a short charter or two on the way home."

Friday at noon Cliff walked into my hangar, check in hand. The 206 was ready to go. I had even topped off the fuel tank. After signing the bill of sale, we watched as his airplane was rolled out onto the apron. After we shook hands, Cliff climbed in, took a look around, and flew away.

Three months later, I heard someone call my name on a street near JC Penney in downtown Anchorage. It was Cliff.

"Hey, Cliff how the hell are you? How's the 206 doing?"

"Real good," he responded.

"No trouble?" I asked

"No, not really."

"What do you mean, 'not really'?"

"Well, I did have one problem."

"Doggone it, Cliff, don't tell me this. You know I would've taken care of any problems with that airplane. You should have called me right away."

"It wasn't the plane," Cliff said. "It was me."

"What do you mean?"

"Well," Cliff said, frowning. "The insurance company said I needed to get ten hours of instruction in the 206 from an FAA-certified instructor before they would insure it."

"What the hell for? That's just nuts."

Cliff told me that the insurance company had required the instruction because he had never flown a plane with a nose wheel rather than a tail wheel.

"That's preposterous," I said. "I've never heard anything that stupid. How many hours do you have in land planes anyway?"

"A little over 33,000 hours," Cliff said.

I shook my head in disgust.

* * *

I got a heads-up from a buddy in Juneau in 1962 about a pilot from the Lower 48 states who had abruptly quit his flying vacation trip to Alaska in his Beechcraft Bonanza 35. Something must have scared him while flying north flying from Seattle—I never heard what—because he dropped to his knees and kissed Mother Earth after stepping out of his airplane in Juneau.

The unhappy pilot threw the keys to a ramp worker at the airport and said, "Tell your boss to sell this thing!" Then he bought a ticket for the next Alaska Airlines flight south.

Within hours, I was on my way to Juneau with a $4,000 cashier's check to buy the Beechcraft. Being halfway to the Lower 48, I decided to continue south to El Monte, California, where Ann and the boys were visiting her parents. We buzzed around Southern California for a week and then headed home.

Richard was about to turn six. First grade awaited him in Anchorage, and he was anxious to get home. He didn't want his best buddy, Brad, who lived across the street, to get a jump on him in first grade. Richard stood up in the back of the plane constantly urging us to fly faster.

We cleared U.S. Customs in Northway and then high-balled it directly to Anchorage in clear weather with ninety miles' visibility. By then, I had the Bonanza figured out. Cleared to land on Runway Six at Merrill, I brought the sweet little bird in slowly, pulling the nose up and setting the main gear down a foot or two beyond the leading edge of the runway.

John Arsenault, the tower operator, called me on the radio. "Way to go," he said. I glanced up. He was giving me a thumbs-up. I guess he liked my short-field landing. Oren Hudson was on a ladder fueling his Grumman Widgeon when I pulled up to the pump and shut down. Richard jumped out and sprinted for the car, determined to beat Brad to first grade.

Oren's eyes were crinkled up in the manner of someone who had been looking into the sun for thousands of flight hours in his Cessna 170B and his Grumman with its powerful Ranger engines.

"Sassara," he declared, "if you can do that in a Beechcraft, I'm going to go buy one."

He wasn't joking, either. Within days Oren left on a buying trip, scouring the country for a good, clean Beechcraft Bonanza. He found one, too, and brought it home to Alaska, where he logged more than 15,000 hours of air-taxi flights on it. He and I became good friends.

Chapter 28
Wolf Meister

Washington, D.C., 1995

Wolf Meister came into my life when his son Christian, my business partner at the time, told me his father was eagerly anticipating his first visit to the United States. Wolf was eighty-four.

"I think you'll like him," Christian said. "He's an old pilot, so you ought to have a lot in common."

As predicted, Wolf and I had a great time sharing stories. Wolf was a delightful man who had a wealth of knowledge and background in aviation. He had flown airplanes I'd only heard about.

Wolf had joined the German army a few years after the end of World War I. Under terms of the Treaty of Versailles, Germany was not allowed to have an air force. To get around this, Wolf and other army enlisted men who passed an aptitude test were sent to Russia to learn to fly old, fabric-covered biplanes. Most did not have tail wheels, just skids. The training was basic.

Wolf worked his way up through the ranks, becoming a highly respected aviator who often was assigned to flight-test new aircraft. He was given the job of flying an experimental twin-engine fighter plane, the new Messerschmitt Me 262. An early version had a tail wheel and was difficult to handle. Later models were equipped with a nose wheel.

World War II had been underway for several years before the new Messerschmitts, each fitted with a pair of jet engines, were supplied to the Luftwaffe. The aircraft flew a hundred miles an hour faster than anything the Allies had.

Wolf also told me a story about Hermann Goering, commander of the Luftwaffe, visiting an air base in Bavaria where the Germans had based one of their new solid-fuel, rocket-powered aircraft which, when launched off its wooden ramp, could soar to 40,000 feet in a flash.

Summoning the pilots, Goering demanded they redouble their efforts for the Fatherland. As the pilots stood listening to Goering go on and on, a contrail appeared overhead—an American bomber. Becoming angry, Goering stomped his feet and pointed up to the contrail.

"Go get the rocket!" Goering shouted, swearing a blue streak.

It was obvious only the rocket-powered aircraft could catch a B-17 bomber flying at 20,000 feet. Being a group leader, Colonel Wolf Meister climbed into the rocket, buckled his seatbelt, and waited.

"There was no throttle or start button," Wolf explained. "The pilot just sat there until someone on the outside ignited the rocket."

"There I was waiting, when all of a sudden I'm on my way, being shot nearly vertical. As I climbed, I pressed the firing button on my guns. Nothing happened. The guns wouldn't work.

"I was streaking straight up at the bomber, but could do little but shoot past it, which I did, coming very close. I looked over at the pilot of the B-17. He looked like a young kid. I'm sure I scared the hell out of him because he rolled the plane into a screaming dive. When I reached the apex of my climb, the procedure was to just glide back down and land.

"When I climbed out of the rocket plane, Goering, a big man, wrapped his arms around me and told everyone, 'That's the way to do it.' From the ground, it appeared to Goering I'd shot down the bomber. I thought it best not to tell him the truth."

I was spellbound by Wolf's incredible stories.

"When I was commander of the Edelweiss Wing in Bavaria, we used a section of the autobahn for a landing strip," he told me. "We rolled the jet fighters under the trees to hide them. After one mission, my crew chief became agitated."

"Colonel, how did you survive?" he asked.

"What do you mean?"

"There are two holes at eye level where a bullet went through the windshield and out the back of the canopy. How come you're not dead?" the crew chief asked.

"Dummkopf!" he said. "Do you think I fly with my head stuck up? Hell no, I am down in the cockpit as deep as I can get."

Later, just as the war came to an end, Wolf crash-landed in a 262 and was hospitalized with several broken bones. "I was lying there in bed bandaged from head to toe when all of a sudden I woke up," he told me. "I smelled coffee, bacon, and toast. 'Hooray,' I shouted. 'The Americans are here.'"

"I was interrogated by a young American lieutenant who thought I was a member of the Nazi party. For weeks I denied this. Finally, I said to him, 'call the British Royal Air Force. They have been trying to kill me for the past four years. They know who I am.'"

"Within hours, an orderly brought me a freshly ironed uniform. The officer of the day, a lieutenant colonel, apologized and told me a Jeep was waiting to drive me home."

For some reason, Wolf was fascinated by the U.S. Congress, so his son asked me to show the former Messerschmitt pilot around the Capitol. As we wandered from office to office in the Senate, I was asked to escort Wolf to the military liaison office. When we got there, the commander of the Air Force and three other generals were waiting. It finally sunk in: Wolf had been the world's *first* jet pilot.

I heard, too, from Tony Lopez, deputy director of the Smithsonian's Air and Space Museum, inviting Wolf to visit. When we arrived, Lopez and a small delegation greeted us with some fanfare. We made our way to an exhibit featuring a fully restored Messerschmitt ME 262. Tony and his staff pulled aside the barricades and carefully pushed a portable stairway up to the wing. Taking off our shoes, we climbed up onto the historic aircraft. Tony slid the canopy back, and Wolf Meister settled into the pilot's seat.

Wolf was like a kid, squirming with joy as he examined the vital instruments and a variety of gadgets. Soon a large crowd had gathered and Wolf began telling stories about flying the 262 in combat. Everyone was thrilled.

Chapter 29
Ann's Favorites

Anchorage, 1980

As Ann flew with me around Alaska, she became familiar with various aircraft, even though she never learned to fly. Without hesitation, she would say her favorites were a De Havilland Beaver or a Cessna 195.

She loved those big, round engines and felt secure sitting in these airplanes. They had an air of purpose. Through the years, I owned and operated four 195s. The first one had a 245-horsepower Jacobs radial. The second and third had 275- and 300-horsepower engines. Last but not least was the turbocharged, 350-horsepower engine that would push the Cessna two-hundred miles an hour at 17,000 feet.

Ann liked the 350-horse best because that aircraft was a special factory-built photo plane in which camera hatches had been installed in the floor and out the baggage door. She would delight in throwing her chewing gum through a camera hole next to her feet. The 195s were a pilot's delight—a gentle aircraft with excellent flight characteristics.

I used to laugh when someone would talk about the lack of visibility forward in the 195. I always told them with a chuckle that you can see a hell a lot through a keyhole if you're close enough.

Once I landed at Opa-locka Airport in Miami looking to get my required biannual check ride from the FAA, which I found operating out of a temporary office in a trailer near the taxiway. After we got the paperwork organized, the check-ride inspector and I walked around the 195. The inspector took his time. Apparently he'd not had a close look at this marvelous aircraft, which becomes more impressive the closer you are to it.

After we buckled in, the inspector looked around the cabin and out the windows while I went through the starting procedure: add a shot of prime, crank the engine, turn over through five blades, and bring on the magneto and distributer. Slowly the powerful engine would come to life, puffing out smoke and then a little flame. Once the engine was settled down and running smoothly, the fun began.

"Can you see out this thing?" the inspector asked.

"Sure," I said, advancing the throttle as we began to taxi away.

"Take me back," he said. "That's enough."

"Don't you want to go up for spin?" I asked.

"No, I've seen enough."

As we taxied back to the FAA trailer, I bragged about the Cessna 195 and what it could do.

"You flew it in here; that showed me you can pass a check ride," the inspector said, then signed my renewal.

Ann's other favorite, the De Havilland Beaver, is a thing of beauty and joy. It was designed as a rugged no-nonsense Bush plane, with military capabilities. The engine is a 450-horsepower Pratt & Whitney 985, with a Hamilton standard propeller. It is somewhat like the Douglas DC-3, pretty hard to wear it out. It is an excellent floatplane, too, and good on skis or wheels. The cabin door was designed to accept a fifty-gallon drum. I added a two-hundred-gallon belly tank for hauling diesel fuel into remote destinations as routine as mining camps and exotic as floating ice islands in the Arctic Ocean.

"Give me a De Havilland Beaver over a private jet any time," Ann said.

Despite not having a pilot's license, Ann had strong opinions about aircraft and pilots, and I learned from experience to listen to them. She had ridden in more than a hundred different models of aircraft, including two rare aircraft—the Funk Model B, a two-seat, single-engine aircraft built in the 1930s and '40s; and the Meyers 200, a hand-built aircraft known for its speed and good looks, built in the 1950s and '60s.

Chapter 30
Mike and Me

Girdwood, Alaska, 1988

Mike Opalka was the Alaska State Trooper in Girdwood, a close community of about 2,000 people. Mike was a big man whose build and demeanor reminded me of the Royal Canadian Mounted Police. For some reason, he and I were drawn to one another.

It was Mike's job to keep the peace in the former gold camp transformed into a ski resort tucked into the Chugach Mountains about forty miles from downtown Anchorage.

Ann and I had bought property in Girdwood in 1980, including a cute little cabin nestled among stands of big trees next to a small stream. We had a good view of the chairlift. Tommy Moe, the Olympic gold and silver medalist, lived next door.

I created a small clearing in front of the cabin for my helicopter so I could commute back and forth to work in Anchorage. Life was beautiful. We had created Alyeska Air Service to cater to tourists and movie companies wanting footage of the mountain and nearby glaciers.

Prince William Sound was a short hop over the mountain. In good weather the flight offered breathtaking scenery, but dangers awaited foolish or unwary pilots when the weather turned nasty.

One day Mike called with news that a Cessna 180 was missing in the area.

"An airline pilot took off from Girdwood headed for Valdez with his family, but didn't show up," he said. "Can you haul me over for a quick look before we start a full-blown search?"

Mike and I headed over to Portage Pass in my Cessna 172. This was an easy flight in good weather. But the pass was only seven-hundred feet high and often high winds blew through it violently when a weather system moved through the mountains.

We had high ceilings and good visibility flying through the pass at about a thousand feet, scanning the steep rock walls on each side. Many lives had been lost here. As we popped out of the pass, I turned left so we could scan the cliff above the railroad tunnel.

Mike grunted. There it was, a 180 smashed into the cliff face about 1,200 feet above the water. It looked like a wad of tinfoil. I circled around and came back in tight, flaps down to twenty degrees, my hand gripping the throttle. As we passed close to the wreckage, I saw what he had seen—a car seat with a tiny baby still strapped into it. At this point my shaken friend, who had seen a lot in his career, broke down in tears.

The rest of the family was scattered around the crash site. The pilot had poked his plane through this tiny notch in the rocks only to discover the Prince William Sound side was shut down by fog. He tried to make a 180-degree turn in a tight space but didn't make it.

Mike Opalka and I often were thrown together on searches or rescues. Sometimes we searched in vain. One day he received a call from the Rescue Coordination Center reporting an airliner passing over the Girdwood area had picked up an ELT (emergency locator transmission). Mike and I flew around for the better part of an hour, listening on our headsets to a strangely constant signal we could not pin down. The signal should have faded as we moved farther from its source. Back in Girdwood, after we shut down the airplane, it suddenly occurred to Mike we should check the ELT in the back of my plane. I crawled back into the tail and pulled it out. Sure enough, it was on. I had to own up to this the following morning over coffee at Peggy's. Nobody cut me any slack.

A week later Ann and I heard screaming when we pulled into the yard. It was the woman who lived two cabins over. "Chuck, come quick, my husband tried to kill himself!"

I sprinted to their cabin and followed her into the kitchen. A man was lying on the floor in a huge pool of blood, and blood was spurting out of his nearly severed left wrist. Jumping into the bloody mess, I grabbed the wrist, squeezed as hard as I could, and told the woman to find a belt to use for a tourniquet to stop the bleeding. The poor man's face was ashen and he was unconscious, but he was still breathing.

Soon Mike arrived, looked at me rolling around on the floor, covered in blood, and said, "We've got to quit meeting like this."

A few minutes later, medics from the Girdwood Volunteer Fire Department arrived and put the suicidal neighbor into a compression bag. Finally I could relax my grip. The man survived.

Not so for a young woman whose life was snuffed out on August 19, 1988, despite a tremendous effort by Mike and the fire department.

The woman, a newlywed, and her equally young husband had taken a Honda three-wheeler for an afternoon ride on the tide flats of Turnagain Arm near Girdwood. The tide in this area is one of the highest in the world, running thirty feet or more. In the shallow arm, during the change of tides, the water races in and out like a fast-moving river. Locals know not to venture out into what appears to be an inviting expanse of tide flats where the mud might support weight when the tide is out but instantly becomes soft and unstable as the tide returns.

The newlyweds were roaring around in circles when they became stuck as the tide turned. The woman jumped off and began pushing. Within seconds, she sank into the quicksand-like mud. In a panic, the husband tried to pull her out while lying across the top of the three-wheeler. He waved his arms frantically at cars passing on the nearby Seward Highway as the water level rose steadily.

When a motorist called 911, Mike and a crew from the fire department rushed to the scene. Having dealt with this kind of emergency before, the firemen brought a portable pump rigged to blast mud away from someone stuck in the flats. By then the water had risen to the woman's knees. Mike was able to free one of her legs while holding her upright and trying to comfort her, but he couldn't wrestle her out of the mud.

"I don't want to die," she screamed in panic.

Tragically, the gas-powered pump could not be started. Everyone was frantic. Even as the tide rose up to his shoulders, Mike lifted the woman with all his might as the tide engulfed her. He kept pulling on the lifeless body until the tide rose over his own head, and he had to give up. A sad team recovered the woman's body during the next low tide.

It took months for Mike to recover from the wound in his heart.

* * *

I'VE SEEN TOO MUCH airplane wreckage in my life. The long list of dead pilots grew longer by the year. The harsh truth is Alaska demands much from those of us who fly. There is little room for mistakes. In good weather, Alaska is the best place in the world to fly. But when weather closes in, a pilot's margin of error becomes ever thinner, pushing the pilot into a dangerous position from which there may be no retreat.

Early on I learned a lesson that has kept me alive. It is simple: don't leave the ground until you *know* you can make it to where you are going. *Thinking* that you can make it is not good enough. The other half of that lesson is, if you are trapped by weather and cannot escape, put the aircraft down in a controlled manner on anything that will support it. Sometimes it's wiser to sustain some damage than to continue flying. I've had arguments with top FAA officials about this.

One time, I encountered a horrendous rainstorm while flying a Super Cub near Jacksonville, Florida. I was ferrying the plane to Alaska for a customer in the Aleutian Islands. The storm was big and black; visibility was going fast, and down below I saw motorists pulling off the Interstate to wait out the storm. Clearly this was no ordinary storm.

I had to find a place to put down the airplane. Lucky for me, a large rectangular, black surface stood out in the rainy landscape ahead. I cut the power and landed. It was a new drag strip, due to open the next day. Two electricians helped me tie down the Super Cub. The men were wiring the lights that controlled the dragsters.

Soon a young woman who worked for the drag-strip operator introduced herself and asked me to meet with her boss, who was on his way. I didn't want any trouble, but saw no reason to refuse. The weather wasn't letting up, and I wasn't going anywhere. A friendly fellow pulled up in his Cadillac. He was brimming with excitement about his idea for an opening-day publicity stunt to promote his new drag strip. Would I stay overnight and race a dragster down the strip the next day?

"You're already all over the news," he said. "I called the radio and television stations and told them about a plane landing on my new drag strip in the middle of a rainstorm."

"Thanks for nothing," I said. "That ought to raise some eyebrows over at the FAA office."

I had hardly gotten the words out of my mouth when I noticed a grim-faced man wearing a white shirt and necktie striding purposely in my direction.

Sure enough, he whipped out a business card identifying himself as head of the FAA flight-standards office in Jacksonville. With no how-do-you-do, this officious gentlemen launched into a tirade citing regulations that I *might* have broken and then lectured me about alternative landing sites that *might* have been available to me had I contacted the FAA.

I listened to him and kept my mouth shut. When he finally finished the tongue-lashing, I stuck out my hand and introduced myself.

"I'm from Alaska and I was taught to put the aircraft down and walk away," I told him. "My mentors all said the same thing: there are repair shops and aircraft parts to fix any damage. That is a much better outcome then cutting a corpse out of the fuselage."

Still outraged, he insisted that I should have called in and gotten a clearance for Jacksonville International Airport. I was fuming by then and lashed back. "That aircraft does not have any instruments. Furthermore, I am not instrument-rated. Crazy bastards like you push pilots into making dumb decisions."

We parted on less-than-friendly terms. I had no regrets about landing in a suitable spot rather than wandering around blindly in the clouds during a violent storm.

Chapter 31
Leaf Cay

The Bahamas, 1992

We dropped the sails of our thirty-seven-foot sailboat, the *Manakai* (Hawaiian for "spirit of the sea"), as we approached the anchorage at Leaf Cay, a beautiful tropical paradise at the north end of the Exuma Islands in the Bahamas. A half-dozen sailboats were settling in for the night.

One boat stood out from among the gleaming white fiberglass hulls. She was a wood-planked hull about forty feet long, a rough-looking boat with a single Mercury outboard hanging low on the transom. The sail rig was the usual Bahama Sloop—a big, loose-footed mainsail set on a short mast and long boom.

As we passed, we waved to the two Bahamian fishermen on board, a young athletic fellow maybe twenty-five years old and an elderly gentleman.

The next morning I pumped up our inflatable and paddled over to invite the two men to share our pancakes and eggs. They accepted our invitation with wide grins. Over breakfast, we discussed the ways of the world, switching from lousy politicians to lousy fishing, and soon we were laughing and having a good time. They shouted their good-byes and an invitation to return to Leaf Cay as they rowed away.

When they got back on board their boat, however, the young man was unable to start the outboard motor, pulling again and again to no avail.

"I'm going to go over there to see if I can get the damn thing started," I told Ann. I had learned to work on outboard motors when I was ten years old growing up in the Panama Canal Zone. Those old 1930s and 1940s motors were famously cantankerous.

Aboard the fishing boat, I noticed the transom was loose and broken. The engine was leaning backward, the head almost touching the water. This didn't make any sense to me until I learned this was a leased boat with fifty percent of the catch going to the owner. I asked for a wrench.

"We don't have any tools," the old man told me.

Returning to our boat, I grabbed a bag of tools, two new spark plugs, and an assortment of bolts, wood screws, and stainless steel ring nails. After I reattached the transom and re-mounted the engine, the rest was easy. The new spark plugs came to life after I disabled the kill switch, which had been shorting out. Without a switch to shut down the engine, I advised the fishermen to allow the engine to idle down and then pull out the choke. This would stop the engine. Soon they were off in search of conch, lobsters, and fish. A section of the old boat's hull had been sealed off and holes had been drilled through the bottom. This allowed seawater to slosh back and forth in a built-in tank where the weekly catch was transported to market. The buyers loved fresh fish.

The Bahamians returned a few days later. As they sailed by, they threw a couple of big lobsters into our cockpit. Some of our fellow yachtsmen yelled at them, waving cash. "These lobsters are already sold," the old man told them.

David, the young fisherman, paddled over to visit. "Hey, Celeste you want to go dancing on the next island? Good music and food." Celeste, our niece, a recent college graduate, did not have to think about it. "Hell, yes, let's go," she said.

David said the old man wanted us to come aboard for dinner. Ann and I grabbed a half-dozen cans of Coke and a bottle of Haitian Barbancourt rum, which in our opinion is the best rum in the world.

It was dark by the time we climbed aboard the old fishing boat. A wood fire was burning in the cut-off bottom of a steel drum on the foredeck. Two cast-iron pots were simmering away. The smell from a spicy mixture of fish, lobster, and conch was wonderful. The old man served us a meal fit for royalty.

The three of us sprawled on the deck looking up at a spectacular display of stars as waves lapped the hull lyrically. We sipped rum and Cokes and talked well into the night. Many times I have told this story, a reminder that a reward of sailing is the people you meet along the way.

* * *

I'D LOVED sailboats for as long as I could remember and Ann had learned to love sailing, too. Little wonder I loved this woman.

We were relaxed and mellow that night at Leaf Cay, caressed by a warm Caribbean breeze. I felt a wave of nostalgia. It felt as if it had been only yesterday—but in fact had been more than forty years—since Ann and I, on our second or third date, strolled the docks at the marina in Long Beach, California. We walked for hours as I explained different hull shapes and sail rigs.

Ann had smiled and nodded as I went on and on.

"Now why is this boat a cutter again," she asked, "that looks like a sloop according to what you said."

"You can spot a cutter because it has two jib sails ahead of the mast," I explained.

"One more time," she said, dutifully. "Show me a yawl."

We found one on the K dock and went over for a closer look. "They are easy to spot because the rear mast is much shorter and positioned behind the tiller or steering wheel," I said.

I told her about Dad buying a twenty-three-foot open sailboat in 1937 when we lived in Miami. Being a sailor most of his life, Dad felt compelled to teach me the art and science of sailing. Later, when I got a pilot's license, I discovered sailboats and airplanes have a lot in common: each uses an air foil for forward movement, in the form of a wing on an aircraft and a sail on a sailboat.

Dad, Mom, Dick, and I used to load the little sloop with groceries and fishing tackle before sailing out Government Cut into the Atlantic. From there, guided by our compass, we crossed the Gulf Stream to Bimini in the Bahamas and thence to Andros Island, the largest of the Bahamian islands. The voyage took four days and nights.

On Andros, we set up a beach camp with our eight-by-eight pyramid tent, gathered firewood for cooking, and proceeded to live a relaxed, carefree barefoot life for a few days.

On one trip Dick and I were playing in the shallows just off the beach when dad hollered at us to get out of the water. I was surprised and a little frightened by the force of his stern words. Two things worried him: a giant black cloud was heading our way, pushed by powerful winds, and a pair of reef sharks were circling around our anchored boat.

As we worked quickly to secure the camp in advance of the fast-moving storm, Dad set out for the water with a hatchet in hand.

"What are you doing?" Mom asked, worry in her voice.

"I'm going out to the boat and check the anchor and make sure the lines are secure," he answered.

"Don't you dare go out there, Chuck," Mom begged. "Please don't do this, I saw those sharks too."

Ignoring her, Dad swam out to the bouncing boat holding the hatchet over his head, ready to defend himself if a shark attacked. He made it back in one piece. I was in awe of Dad.

Chapter 32
Good-bye, Ted

Anchorage, 2004

I was stunned and saddened when my friend Ted Stevens died in a crash while flying to a private fishing lodge in Southwest Alaska. I reflected on his long service in the U.S. Senate and thought back almost fifty years to our days as brash, young Alaska legislators charting the future of the new state.

Ted had been aboard a ten-passenger, single-engine de Havilland Otter on floats that slammed into a mountainside on its way back to the lodge after a day of fishing. Cause of the crash was not known, but in the aviation community there was speculation that one factor might have been the "have-to-get-home" syndrome.

Five of eight people aboard were killed. Ted had been sitting directly behind the pilot, a former airline pilot with a wealth of experience in Alaska.

Ted's party had overnighted at the lodge and had flown out to fish the next day. Late in the day, the weather was marginal—low ceilings and limited visibility. But the lodge was just a short hop away. The pilot might have thought about waiting on some lake or river for improved conditions, but there had to have been pressure on him to get his weary passengers back to the comforts of the lodge. A decision was made to fly.

According to first responders, both of the pilot's arms were found in his lap. It appeared he may not been holding the control wheel when the Otter hit. Significantly, neither of the pilot's arms were broken, which is commonly found in crashes in which a pilot is gripping the wheel tightly on impact. This led to speculation the pilot had been stricken somehow and might even have been dead before the impact.

Many of us wondered if Ted, had he been in the front-right passenger seat, might have been able to take over the controls and avoid the crash—if, in fact, the pilot was incapacitated. Having been a pilot in World War

II, Ted's reaction would have been to grab the wheel and pull up the nose. Even if unable to clear the ridge, he might have been able to pancake the plane onto some trees. The floats would have absorbed some of the impact.

Ted had logged many hours for the Army Air Corps during the war, flying supplies from India into China. He and I shared many flights over the years, and I often turned the controls over to him. He appreciated this courtesy and showed me he had not lost it. Flying is like riding a bike. You don't forget how to do it, you just get rusty. I am confident Ted would have had the skills necessary to bring the plane back under control and land it safely.

Despite an exhaustive investigation, including scores of interviews, the NTSB was unable to solve the mystery of exactly what happened on an August afternoon in 2010, reporting only that the pilot had been temporarily unresponsive "for reasons that could not be established."

* * *

TED'S DEATH in a crash brought back memories of the tragic loss of his first wife, Ann, in December, 1978.

Ann Stevens and three others were killed and Ted was seriously injured in a freak accident at Anchorage International Airport. A corporate-chartered Learjet 25C carrying six passengers was just about to touch down on Runway Six when a sudden cross-wind flipped the plane onto its back. The only other survivor besides Ted was Tony Motley, who later became U.S. ambassador to Brazil.

It was a sorrowful crowd that gathered to share Ted's grief at the family's modest log home in Anchorage. Ann Sassara and Matilda Stepovich, wife of the last territorial governor, stepped in to hold things together at the house and care for the five children. My Ann and Ann Stevens were close friends who loved to share stories about their grandchildren.

I was on our sailboat in Honduras when the crash occurred. When Ann called, I rushed back to Anchorage. After Ted was released from the hospital, where he was treated for head, neck, and arm injuries, he and I drove around Anchorage for several days discussing the direction of our lives. It was a sad, somber time.

One night a few weeks later Ted asked me to take him back to the hospital. "I've got to pick up my brain scans," he said.

After leaving the hospital, we stopped the car under a streetlight, and Ted pulled out his x-rays. After studying them at great length, we concluded there was nothing to be found inside of Ted's thick skull. We chuckled over this little joke and then laughed uproariously when Ted swore me to secrecy.

"Don't you tell anyone we couldn't find my brain," he said.

Hearing Ted Stevens laugh again was music to my ears.

Chapter 33
The Friday Professionals

Anchorage, 2014

I was always a little late arriving for the Friday-evening gathering of pilots in Bill Hately's pristine hangar on the shore of Lake Hood, the largest floatplane base in the world.

The noise level on summer evenings was sometimes deafening as De Havilland Beavers; Cessna 180s, 185s, and 206s; hundreds of Piper variants, and an assortment of other light floatplanes climbed up onto the step and roared away.

The activity was precisely controlled by a dedicated control-tower crew. Planes on floats lined the lakeshore; other aircraft with conventional landing gear, mostly tail-draggers, were tied down on every square inch of the property as far as the eye can see. This was truly an amazing place.

The pilots arrived in Bill's hangar with food and drink, including pizza, smoked salmon, and hot and cold dips. I usually brought fresh fruit in season or some weird off-brand beer.

At ninety-two, Oren Hudson was a respected elder in this group. His first flight was in a Taylor J-2 Cub in 1938. Two years later he hired on as a copilot for Trans World Airlines (TWA), flying Douglas DC-3s, and over his life of flying logged 37,666 hours of flying time.

This evening Hudson reminisced about his experiences flying Stinson Detroiters and other old round-engine, fabric-covered aircraft. His companion were discussing the ins and outs of flying the Noorduyn Norseman on the rivers of Alaska in the 1930s and early '40s. The Norseman was designed and built in Canada as a Bush plane on floats—a hugely successful, non-nonsense aircraft with a 600-horsepower Pratt and Whitney engine.

Reinhold Thiele told a story about the time he was flying south along the Kuskokwim River in a Norseman when he encountered Oren Hudson roaring in the opposite direction in a Stinson SR-JR, another big, old Bush plane with a radial engine. The two pilots landed on the river, beached their planes, and proceeded to discuss flying conditions and perhaps trade a loaf of bread for an onion.

Our host was Capt. Bill Hately, a retired airline pilot born and raised in Bethel, who insisted I accompany him to an adjacent hangar for a look at his Grumman Widgeon. There it sat, one of the most beautiful twin-engine amphibians ever built. Bill had completely refurbished and painted this stunning airplane. Twice in my life I had walked into a bank with a check in my hand prepared to buy a Widgeon, only to discover some other pilot had gotten there ahead of me. The first time this happened was a day before the 1964 earthquake. Given the demand for aircraft in the earthquake recovery effort, I would have hit it big on this deal if I had gotten there five minutes sooner.

Meanwhile, in the middle of this evening of animated story-telling, Alex McInroy walks in, and a shout goes up. All at once we're firing questions at him—where had he been? What had he been doing? None of us had seen Alex in years. The pilots pumped Alex's hand and hugged him. He looked great, still slender as a willow branch at eighty-eight. I managed to pry him loose and wrapped my arms around him.

"Remember the time I ran out onto the ice on Lake Hood to see if you were still alive when your student stalled out and slammed back down on the ice, then pissed his pants?" I asked him.

"Ya, ya, this was fifty years ago, and I think the statute of limitations has run out on that story," Alex said, laughing.

Raised in Anchorage, Alex became a Bush pilot and flew just about everything there was to fly in the 1940s. When I met him, he was at the controls of a Short Skyvan, a turbine-powered box car built in Ireland. He retired at sixty-five as a commercial jet pilot.

Later Ron Sheardown and I were munching boiled shrimp and strawberries, sharing stories. Ron is our resident Canuck, a Bush pilot from Canada who migrated to Alaska in the early 1960s. He spent most of his flying career in the Arctic, where he had made several extraordinary rescues and flights over the North Pole. Once he landed his ski-equipped Cessna next to two men huddled behind their tipped-over snowmachine trying to survive a seventy-knot wind at minus forty degrees Fahrenheit. Visibility was near zero right down to the ground. But he knew where the two men were and slowly let down against the fierce wind.

The wind was so strong that Ron, who described himself as a self-taught helicopter pilot, was able to stop his forward motion in mid-air and, by slowly reducing the power, could fly backward! He eased down to the icy surface, touching his skis right next to the two men—an unbelievable test of skill and courage.

Ron once bought a Russian-designed Antonov AN-2, a huge 1,000-horsepower, single-engine bi-plane built in Poland. Using a blank tablet and a lot of pencil lead, he navigated his new airplane from Poland over the North Pole to Barrow, Alaska.

A few years later, Ron flew to the North Pole, accompanied by another pilot flying a Cessna 185 on skis. When they reached the pole, Ron landed the AN-2 and got out to look around. The 185 landed nearby. The big Antonov was on wheels but, as expected, the extreme cold and wind had packed the snow down to a hard surface Ron thought would be suitable for a wheeled landing. To Ron's dismay, however, the wheels of the AN-2 began to settle into the surface. Within minutes, the landing gear had disappeared, and the plane was resting on the bottom set of wings. There was nothing the two pilots could do. Ron recalculated his position, confirming his plane was stuck at exactly the top of the world.

As much as they hated leaving Ron's plane, they had no choice but to head south to Alaska, planning to return with a recovery team.

They never saw the airplane again. Because the ice constantly moves, and pieces break off and slide over each other, the plane's resting place had become unstable before their return. The old bird fell through the ice and likely sank to the ocean bottom.

Another arrival was Fred Walatka, whose usual contribution of smoked salmon strips was devoured quickly. This reminded me that when I flew a mail route on the Yukon River, I always stopped in at the village of Holy Cross to pick up a supply of squaw candy, as it was indelicately called. I must have smelled fishy afterward because many of my passengers would ask me to vent a little fresh air into the cabin.

Freddy is a second-generation pilot from Bristol Bay whose dad I knew back in the 1950s. Freddy flew a 65-horsepower Piper Cub on floats and a Stinson SR-JR, a big, old tail-dragger built in 1929 and powered with a 300-horsepower radial.

The last to come aboard was Boyuk Ryan from Unalakleet, another second-generation pilot who became president of Ryan Air. Boyuk and his brother Adrian both worked for their dad fueling and maintaining airplanes. However, Adrian somehow slipped through a crack, trading the good life of aviation for that of a respected orthopedic surgeon in Anchorage.

The days of the old Bush pilots were fading. These gatherings brought back memories of fire-potting frozen engines; digging an ax out of your plane to hew a new tail skid from a spruce bough; and sitting out bad weather alone in the middle of nowhere.

The eager faces of the Friday crowd were wind-burned and mapped with age, but the pilots' eyes were snapping bright and lit from within. At an age when most octogenarians discuss their physical ailments, this group still exuded the fire of life, still ready to challenge the elements and soar into the clouds.

It was noted by Oren Hudson that all eight pilots in attendance this evening had flown Grumman Widgeons. Then we got to talking about the Republic Seabees, which had a short but colorful history along the Kuskokwim and Yukon rivers. Somebody remembered the time Ray Petersen sent a new pilot out to Port Alsworth on Lake Clark. Someone waved to the pilot warning him that his landing gear wasn't down as he lined up to land on the airstrip, so the pilot went around, pumped his gear down, and for unknown reasons landed on the lake. The wreckage sat on the lake bottom for years.

I had my own Seabee story. One day I was holding court in my office at Merrill Field when I heard the squishing sound of water-soaked shoes. It was my good buddy and fishing partner, "Big Bill" Smith. He was dripping wet, carrying a briefcase in one hand and an expensive shotgun in the other.

"What to hell happened to you?" I asked.

"You know those damn Seabees," Bill exclaimed. "I bought one down in Seattle and flew it all the way up here, then it stopped running three miles from home. I had to land it in Turnagain Arm. I drifted around for an hour until it bumped into a sandbar, where I managed to climb out. Then the tide took it out again."

"How did you get ashore?" I asked.

"An Air Force helicopter saw me standing out there and picked me up."

"What are you going to do with the plane?"

"Aw, I called George Grant and told him he could have it cheap. My luck has run out with Seabees. Three in a row have quit. I should have stayed with my original amphibian Douglas Dolphin," Bill said.

I remembered Bill had sold the Dolphin to somebody in Long Beach, California, who flew movie stars and celebrities among others to Catalina Island.

After I told this story, Bill's son Jack wanted me to tell about the time his dad was arrested. I was glad to oblige. Bill and I had leased a thirty-two-foot Bryant Gill Netter and planned to fish out of Clark's Point in Southwest Alaska.

I was staying on board in Dillingham, keeping an eye on things, when a kid came running down the dock. "Come quick," he said. "Bill Smith is in jail."

I followed the kid up to magistrate's office to find out what was going on. The timing of this crisis could not have been worse because the commercial fishing season was scheduled to begin the next morning. I was shocked to learn an agent of the state Fish and Game Department, being the closest thing to law enforcement, had put Bill behind bars on a complaint of attempted murder.

There was no time to argue about Big Bill's guilt. We had to go fishing. So I asked the agent to release my fishing partner from jail on his own recognizance. He refused, so I borrowed his phone to call state Attorney General Avrum Gross in Juneau.

I explained the situation to Av as best I could. Apparently my partner was behind on the insurance payments for his Cessna 180, and a banker was seen breaking into the plane, apparently intending to repossess it. Somebody tipped off Bill, who raced up to the airstrip to save his Cessna. Finding the banker inside the plane, Bill reached in under the back seat, pulled out a rifle, and began loading a shell in the chamber. The banker jumped out and ran straight for the magistrate's office, where he filed a warrant for Bill's arrest. Bill was picked up on the dock.

When the banker returned to the airstrip, he found the Cessna 180 sitting on its gear legs, stuck in the mud—no wheels, no propeller, no radios. Word of Bill's troubles had quickly circulated through Dillingham. A friend had pulled off the parts and hidden them in the village. The plane wasn't going anywhere.

The following morning, Bill and I went fishing. I was responsible for getting Bill to his trial a month later. He was found not guilty.

Once the stories began flowing, there was no stopping us. Something someone said reminded me of the winter day, sixty years after learning Santa came to Miami by air, that I landed my Cessna 206 on the hard-packed snow alongside one of the string of villages along the Bering Sea Coast. The plane was packed full of Christmas presents mailed from all over the world. Excited villagers met me with dog teams and snow machines. The northern lights were weaving their mysterious patterns across the night sky. Who would have ever thought I would fulfill my destiny here?

I was Santa Claus!

These rich memories of a lifetime of flying were more valuable than gold. I recalled an evening many years ago flying a Piper PA-14 northbound through the Yukon Territory when bad weather and darkness forced me to land on a sandbar next to a river in the middle of nowhere. In the back of the plane I found a newspaper, which I spread out on the ground for insulation. Then I unrolled my sleeping bag, built a fire, and toasted a three-day-old cheese sandwich, listening to the wolves exchange messages across the Arctic wilderness.

Life doesn't get any better than this, I thought to myself as I settled down for the night.

Author's Notes

While researching and writing this book it was a challenge to plumb my mind recalling stories and characteristics of the many airplanes I have known. For fellow pilots and armchair pilots as well, I offer a list of aircraft I flew between 1951 and 2013. I have come to think of this collection of airplanes as a hobby gone wild.

Aero Commander 100: 150 horsepower, single engine.

Aero Commander 200 Darter: 180 horsepower, single engine.

Aero Commander 520

Aero Commander 560: I put 1,600 hours in this aircraft; it had the best single-engine performance of all the Aero Commanders and ran on eighty-octane gas.

Aero Commander 680

Aeronca 7AC Champ

Aeronca 11AC Chief: I had one equipped with a Johnson bar mounted inside the cabin that was hooked to a cable that went out to the prop hub. It was an inside rewind starter that worked like a starter on an outboard motor. After taking it apart and winding the flat spring up again for the twentieth time I took it off the airplane and threw it away, then went back to hand-propping the way the Lord intended!

Aeronca 15AC Sedan: This was the plane my mother bought. It was good on wheels, skis, or Edo 2000 floats. I learned dead-reckoning navigation in one while flying six-hundred hours all over Alaska with no radios In later years I used what I had   learned in that Sedan to fly a broken-down Lockheed Constellation thousands of miles over jungle to the head waters of the Amazon river.

Aeronca ACA

Aeronca DC

Aeronca ECA

Aeronca L3 Grasshopper

Champion Aircraft's offering to the gods: First they built a twin-engine version of the proven two-place "Champ" that the manufacturer expected to become the darling of the flight schools. It was placed on tricycle gear. Two 100-horsepower engines were mounted on the wings. Phony switches were added to simulate retractable gear, flaps, etc. The regular Champ cruised 100 mph on five gallons of gas. The new twin engine Lancer cruised at 100 mph on ten gallons of gas. The whole idea was dropped.

Then Champion built a tri-gear version, which was presumed to be easier to fly. It had a steering wheel in front and a stick in the rear. It felt strange flying that little tail-dragger with a steering wheel. But wait! Champion wasn't finished. A flaming redhead my brother was dating bought a new Champion, a DXer, and asked him to look it over. He called me. "Hey, Chuck, how about we meet at Merrill Field, you fly it, and see what you think?" It turned out to be an aeronautical engineer's wet dream. It was a Champ all right, sitting proudly on tricycle gear mounted backwards! The main gear was mounted where it should be for a tail-wheel version but there was a nose strut sticking out of the belly with a matching 600x6 tire and wheel. What a mess. I screwed up my courage and rolled down the runway. When I pulled back the control stick, I rocked back on the single wheel and rolled away until we were airborne. Now came the real fun. When I landed in my usual three-point attitude the *belly* wheel touched. I had the sensation of zooming down the runway like I was on a pogo stick. Suddenly it stalled out and we flopped forward onto the main gear. By then I had enough of that plane. After some dual instruction the Redhead flipped over while landing on a sand bar. The plane was a total loss.

Alon Ercoup: Features include a sliding bubble canopy and a 90-horsepower engine.

Beechcraft Musketeer 23

Beechcraft 35 Bonanza: The first one I owned had a Beech electric prop, which worked great. A series of later models had bigger engines, heavier spars, and constant speed props.

Beechcraft Debonair 33

Beechcraft Bonanza A36

Beechcraft Twin Bonanza C50

Beechcraft Baron 58

Beechcraft Travel Air 95

Beechcraft Sport 150

Beechcraft Sierra 180

Beechcraft Sierra 200

Beechcraft Skipper

Bell 47: I found this helicopter easy to fly having spent a hundred hours in the Brantley B2B (see below).

Bellanca Citabria: This aircraft is essentially an upgraded Aeronca Champ.

Bellanca Cruisair 14-13: 150 horsepower. While I was in Florida I worked with Al Nickerson, who never met a Bellanca he did not love. Every time a new model came out, he had to have one. So he would sell me his old Bellanca, which I resold in a matter of a few weeks.

Bellanca Cruise Master 14-19: 165 horsepower.

Bellanca Cruise Master 14-19: 190 horsepower, tri-gear.

Bellanca Cruise Master: 230 horsepower, Continental engine, tail wheel.

Bellanca Cruise Master: 260 horsepower, tri-gear.

Bellanca Viking: 300 horsepower.

Britten Norman Islander: This is a fixed-gear STOL twin powered by two 250-horsepower Lycoming engines. Both the stall speed and the single-engine control speed were about 35 knots.

Brantley B2B: Flying the Brantley, in which I soloed after five hours' flying, represented my first venture into choppers. One day I decided to move it over to the gas pumps from a gravel pad. I flew it off the ground, over the top of a bunch of parked planes, and around the hangar setting it down next to the gas pumps. Joe Wilbur, who was teaching me to fly it, asked how I had moved it. "I flew it over here but I didn't go very high," I told him. Laughing, he threw his hands in the air and said, "Get outta here with that thing!" I passed the solo test.

Bowers Fly Baby: Mostly a wooden home-built with folding wings.

Call Air A2 Interstate Cadet: 125 horsepower.

Arctic Tern: This was a copy of a Call Air built in Alaska by Bill Diehl. I sold Bill his first Interstate Cadet.

Cessna Single Engine

Cessna 120

Cessna 140

Cessna 140A

Cessna 150

Cessna 150 Aerobat

Cessna 152

Cessna 170

Cessna 170A

Cessna 170B

Cessna 171: nose-wheel conversion, very ungainly.

Cessna 172

Cessna 172: 195 horsepower.

Cessna 175

Cessna 177

Cessna 177 RG: I sold five of these in one year!

Cessna 180: This was a fantastic airplane. It could do everything well on wheels, floats, or skis.

Cessna 182

Cessna 182RG

Cessna 185: a 180 on steroids.

Cessna 185 Amphibian

Cessna 190: 240-horsepower Continental engine.

Cessna 195: 245-horsepower Jacobs.

Cessna 195: 275-horsepower Jacobs.

Cessna 195: turbocharged 350-horsepower Jacobs. Now you're talking! I loved passing retractables with my landing gear hanging down in the breeze.

Cessna 205

Cessna 206: This may be the *best* all-around light plane ever built.

Cessna 207: This was a flying school bus but you couldn't climb much above 9,000 feet with a big load.

Cessna 210

Cessna 210: turbocharged, pressurized.

Cessna 310: I put a lot of time in a Q model, which was fast and efficient.

Cessna 320 Skyknight

Cessna 336 Skymaster: Keep a wary eye on the rear prop.

Cessna 401: I liked the 401 and 402 airplanes but not the Continental engines. I dealt with cracked cylinders in cold weather.

Cessna 402

Cessna UC78 Bobcat: WW II twin-engine trainer.

Commonwealth Sky Ranger: This was a homely mutt built in the 1940s.

Curtis Commando C46: On my first flight, I was in the left seat and lost both engines due to fuel contamination and had to belly it in to a small farmer's field.

Culver V

De Havilland Beaver

De Havilland Dove: A handsome twin with inverted six-cylinder engines and air starters.

Dornier DO27: Tough, single-engine with a 260 Lycoming GO435.

Dornier DO28: Twin with 250-horsepower Lycoming engines mounted in pods, rearward-facing; middle seats.

Emigh Trojan: This was an odd four-place aircraft, all metal with the reinforcing ribs on the outside of the skin.

Enstrom Helicopter: The rubber-band drive bothered me but seemed to work.

Ercoupe 415 C

Ercoupe 415 D

Ercoupe 415 E

Fairchild 24: 165-horsepower Warner Radial. This was a sweet-flying plane with roll-down windows. I flew one from Seattle to California to Florida with my elbow sticking out the window.

Fornair: Another version of an Ercoupe.

Funk Model BC: Harry Funk and his brother built these planes using Ford Model B engines with pressurized crankshafts. The old CAA came out with an airworthiness directive requiring that engines be overhauled every fifty hours. So the brothers changed over to 85-horsepower Continental engines, exempting themselves from the requirement. The "C" in the model name designed Continentals.

Globe Swift: I owned three and loved every one.

Grumman Widgeon: This was truly a great plane. I flew an original Ranger-powered model equipped with Curtis Reed fixed-pitch props, but twice I came within minutes of buying versions with Lycoming GO435-equipped engines.

Helio Courier: I considered this aircraft nearly worthless for Bush flying. It was practically impossible to climb up onto the plane to refuel it with a five-gallon can of gasoline. Also, the tail was too heavy for a pilot to pick up and move around, though it performed well on nice paved runways. Our Bread and Butter planes all had dents and finger prints permanently squeezed into every surface as we hung on to our planes when we were alone.

We fastened steps and handles all over our Cessnas, plus we had ropes dangling from the wing struts and dragging along behind the tail. In my case, I had a thirty-five-foot piece coiled up under my seat and of course a pair of twenty-foot lines attached to the bow of each float. Like the Boy Scouts say, "be prepared." When you are standing alone on a sandbar next to a fast-moving river and have to hand-prop a three-hundred-horsepower, fuel-injected engine mounted on a Cessna 206, you realize it's like first-time sex—you kind of know what to do, but you're not so certain about the outcome!

Howard 250: This WW II bomber conversion was one fast airplane.

Lake LA4 Amphibian: 180 horsepower.

Lake LA4 Amphibian: 200 horsepower: This was a nice-flying sport plane best operated on smooth water.

Lockheed LASA 60: A bush plane designed by Lockheed engineers who lived and worked in Burbank, California. It was a rugged-looking bird with swept-back gear legs, a massive nose gear, huge cargo door, and a turbocharged IO 470 Continental engine. I bought one and flew it from Seattle to Florida, then turned right around and flew it back to Bethel, trying to figure out how to make money with it. The plane couldn't climb worth a nickel. The other problem was the balance, which was such that the more weight you loaded in, the more it pushed down onto the nose gear. A guy in Kenai hired a pilot from South Africa who thought the Lasa 60 was a great airplane and talked his boss in Kenai into trading a Cessna 206 to me for the Lasa. That deal made me jump for joy.

Luscombe 8A: two place, all metal, 65 horsepower.

Luscombe Sedan: This was a handsome four-place, all-metal tail-dragger with 165-horsepower Continental.

Maule M4: 145-horsepower Continental.

Maule M5: 210 Continental.

Maule M5: 215-horsepower Franklin, then an O235 Lycoming.

Mooney Model 20

Mooney Model 21

Mooney Model 231

Morrisy 100: This plane was designed by an engineer at Boeing as an all-metal tandem seat sport plane. Newer versions were manufactured by George Varga in Arizona.

North American Navion: 205 horsepower.

North American Navion: 225 horsepower.

North American Navion: 260 horsepower. I owned five of these big old rocking chairs, which were a little slow but sturdy, all in all a nice roomy flying machine.

Pilatus Porter land plane

Pilatus Porter on floats: Built in Switzerland, these planes had a stick, a rudder bar, a bomb bay, and a set of rails inside the tail to secure passenger seats when it was not carrying freight. The CIA tried to hire me to fly them in Vietnam. I declined because I had two boys to raise.

Piper J-3 Cub

Piper J-4 Coupe: two place, side by side.

Piper J-5 Cruiser: This later became the Piper PA12 Super Cruiser.

Piper PA-11

Piper PA-12

Piper PA-14: This was basically a four-place Super Cub. My guess is that nearly all that were built are in Alaska today.

Piper PA-15 Vagabond: This little 65-horsepower plane with two-place, side-by-side seating, control sticks, and no shocks was a bare-bones little tail-dragger. My brother Dick bought one to fly fresh salmon off the beach near Kenai into Anchorage, where he resold the fish for a nice profit.

Piper PA-16 Clipper: This was a four-place, short wing, 108-horse-power plane with no flaps and a control stick on 1400 EDO floats. This baby took talent to get into the air.

Piper PA -17 Vagabond: It had bungee-cord shocks.

Piper PA-18 Super Cub: 90 horsepower.

Piper PA-18 Super Cub: 135 horsepower.

Piper PA-18 Super Cub: 150 horsepower.

Piper PA-20 Pacer

Piper PA-20 float plane

Piper PA-22 Tri Pacer: This aircraft is under-appreciated by pilots who have never flown one but have two-bit opinions nonetheless.

Piper PA-22 float plane

Piper PA-23 Apache: twin engines, 150 horses. The Apache could maintain 2,000 to 3,000 feet on one engine. It was designed by Stinson, then sold to Piper Company; the last vestige of Stinson used on Pipers was the overhead crank for elevator trim.

Piper PA-23 Aztec: In 1960, I took the first one I ever flew over the top of Mount McKinley at 23,000.

Piper PA-24 Comanche: 180 horsepower.

Piper PA-24 Comanche: 250 horsepower.

Piper Pawnee: 150-horsepower spray plane.

Piper Pawnee: 235-horsepower spray plane. Both the 150 and 235 were agricultural airplanes that flew like trucks, which they were!

Piper PA-28 Cherokee 140

Piper PA-28 Cherokee 150: I shook hands with William T. Piper in Anchorage when Piper introduced its new all-metal plane Cherokee 150 to the world in 1960.

Piper PA-28 Cherokee 160

Piper PA-28 Cherokee 180

Piper PA-28 Cherokee 235: It could lift its own weight.

Piper PA-28R Arrow: 200 horsepower.

Piper PA-28R Arrow: turbo-charged.

Piper PA-28 Cherokee 180 on floats: She flew okay but did not handle well near docks, next to steep shorelines, or around commercial fishing boats.

Piper PA-31-310 Navajo

Piper PA-31-350 Navajo Chieftain: In 28,000 hours of flying, this was the only airplane that I went to school for. I attended a flight-safety course in Lakeland, Florida for four days to learn its systems. But I lost my patience when I discovered I would have to wait another two days for a flight check. Asking politely which Chieftain on the apron was mine, I left the director of the school standing with his jaw hanging open when I climbed in and flew away to Anchorage. I would rate this model higher than the Cessna 400 series because of its robust Lycoming engines.

Piper PA-32 Cherokee 260: seven place.

Piper PA-32 Cherokee 300: The 300 was a seven-place work-horse with the great Lycoming IO 540 engine.

Piper PA-32R Lance: This model was a retractable-gear version of the Cherokee 6 with 300 horsepower.

Piper PA-34 Seneca

Piper PA-34 Seneca II

Piper PA-34 Seneca III

Piper PA-38 Tomahawk: This was a T-tailed trainer built to compete with the Cessna 150 and 152 trainers.

Piper PA-44 Seminole: A nice little T-tailed, four-place aircraft.

Piper T-Tail Lance: Horrible Mistake! The stabilizer was located at the top of the rudder. A factory pilot flew one to Alaska to show me. I flew it around the field three times and told the Piper pilot he might as well take it back south with him. No one was going to want an aircraft in which you could not lift the nose gear until the plane reached 60 miles an hour. You had to get off the ground more quickly in Alaska.

Rallye Minerva: An aircraft importer in Bartow, Florida called me with the news he was uncrating a new plane that would jump over tall buildings and land on a dime. Curious, I flew my Cessna 180 over for a look at this wonder. When I looked it over, I had to laugh. The importer wanted to know what was so funny.

"You can't roll a fifty-gallon drum over that wing and up over the side of the airplane, then topple it down to the front seat. Or tie cargo on the outside of the thing."

"Why would you want to do that?" he asked.

"Well," I answered, "Hauling fifty-gallon drums of diesel is what Bush pilots do for a living."

Republic Seabee: I had two of them, both under-powered, but they handled rough water well.

Robinson R-22 helicopter: It was a shock the first time at the controls when I pulled it off the ground and went vertical forty feet to clear some trees.

Stinson 10 A: 90-horsepower Franklin engine.

Stinson Voyager 108: 150-horsepower Franklin.

Stinson Station Wagon 108-2: 165-horsepower Franklin.

Stinson Station Wagon 108-3: 165-horsepower Franklin. This aircraft had a big tail and fifty-gallon fuel tanks.

Stinson Station Wagon 108-3: 190-horsepower Lycoming with a controllable-pitch prop.

All Stinsons on EDO floats.

Stinson L5: I flew this airplane on Wallam wooden floats hand-made from plywood. I had been called in to test hop the plane at Lake Hood. Big mistake! Its useful load was limited to three-hundred twenty-five pounds, which amounted to one fat pilot and about six gallons of fuel. After I finished there was an entry in the log book that they had to live with instead of just using a pencil to subtract the weight of the landing gear, then add the estimated weight of the floats.

Taylorcraft BC-12D: This was the poor man's Super Cub. These planes run around in a tiny circle then up on one float and you're gone while the guy in the Cessna 180 is still on the water. I did my buddy Chuck Higgins a favor when I delivered his 1938 Model 12 with an uncowled engine, big round steering wheel, massive tachometer, and a single nose tank with twelve gallons of fuel on board. I flew from Anchorage down the length of the Kenai Peninsula and across the Gulf of Alaska to Kodiak Island at 80 mph. Believe me, I was happy to set those little Taylorcraft tires on the ground.

Thurston Teal: This was a nice, two-place amphibian with a forward-facing engine mounted on a pylon with the prop whirling around right over your head.

Varga Kachina: 180 horsepower. This was built by Morrisey Aviation in Arizona, featuring all push-rod controls and an outer skin that was thirty-two thousandth of an inch thick. It climbed like mad, and it could twist and turn and dive like a World War I fighter plane.

Varga Kachina, tail-wheel conversion: George Varga paid my way to Payson, Arizona, to demonstrate this plane to an Israeli Air force General. It looked like a baby North American T-6. All that was missing were some machine guns.